KENT

118 Street Plans of town centres and villages
Road Maps with index
Population Gazetteer
Administrative Districts
Postcode Districts
Index to street plans

Maps prepared and published by
ESTATE PUBLICATIONS
Bridewell House, Tenterden, Kent
and based upon the Ordnance Survey maps
with the sanction of the controller of H.M. Stationery Office

The publishers acknowledge the co-operation of local authorities
of towns represented in this atlas.

ESTATE PUBLICATIONS

STREET ATLASES

ASHFORD, TENTERDEN
BASILDON, BRENTWOOD
BASINGSTOKE, ANDOVER
BOURNEMOUTH, POOLE, CHRISTCHURCH
BRIGHTON, LEWES, NEWHAVEN, SEAFORD
BROMLEY, (London Borough)
CHELMSFORD, BRAINTREE, MALDON, WITHAM
CHICHESTER, BOGNOR REGIS
COLCHESTER, CLACTON
CRAWLEY & MID SUSSEX
DERBY, HEANOR, CASTLE DONINGTON
EDINBURGH
FAREHAM, GOSPORT
FOLKESTONE, DOVER, DEAL
GLOUCESTER, CHELTENHAM
GRAVESEND, DARTFORD
GUILDFORD, WOKING
HASTINGS, EASTBOURNE, HAILSHAM
HIGH WYCOMBE
I. OF WIGHT TOWNS
LEICESTER
MAIDSTONE
MEDWAY, GILLINGHAM
NOTTINGHAM, EASTWOOD, HUCKNALL, ILKESTON
PLYMOUTH, IVYBRIDGE, SALTASH, TORPOINT
PORTSMOUTH, HAVANT
READING
REIGATE, BANSTEAD, REDHILL
RYE & ROMNEY MARSH
ST. ALBANS, WELWYN, HATFIELD
SALISBURY, AMESBURY, WILTON
SEVENOAKS
SHREWSBURY
SLOUGH, MAIDENHEAD
SOUTHAMPTON, EASTLEIGH
SOUTHEND-ON-SEA
SWALE (Sittingbourne, Faversham, I. of Sheppey)
SWINDON
TELFORD
THANET, CANTERBURY, HERNE BAY, WHITSTABLE
TORBAY
TUNBRIDGE WELLS, TONBRIDGE, CROWBOROUGH
WINCHESTER, NEW ALRESFORD
WORTHING, LITTLEHAMPTON, ARUNDEL

COUNTY ATLASES

AVON & SOMERSET
BERKSHIRE
CHESHIRE
CORNWALL
DEVON
DORSET
ESSEX
HAMPSHIRE
HERTFORDSHIRE
KENT (64pp)
KENT (128pp)
SHROPSHIRE
SURREY
SUSSEX (64pp)
SUSSEX (128pp)
WILTSHIRE

LEISURE MAPS

SOUTH EAST (1:200,000)
KENT & EAST SUSSEX (1:150,000)
SURREY & SUSSEX (1:150,000)
SOUTHERN ENGLAND (1:200,000)
ISLE OF WIGHT (1:50,000)
WESSEX (1:200,000)
DEVON & CORNWALL (1:200,000)
CORNWALL (1:180,000)
DEVON (1:200,000)
DARTMOOR & SOUTH DEVON COAST (1:100,000)
GREATER LONDON (1:80,000)
EAST ANGLIA (1:250,000)
THAMES & CHILTERNS (1:200,000)
COTSWOLDS & WYEDEAN (1:200,000)
HEART OF ENGLAND (1:250,000)
WALES (1:250,000)
THE SHIRES OF MIDDLE ENGLAND (1:250,000)
SHROPSHIRE, STAFFORDSHIRE (1:200,000)
SNOWDONIA (1:125,000)
YORKSHIRE & HUMBERSIDE (1:250,000)
YORKSHIRE DALES (1:125,000)
NORTH YORK MOORS (1:125,000)
NORTH WEST ENGLAND (1:200,000)
ISLE OF MAN (1:60,000)
NORTH PENNINES & LAKES (1:200,000)
LAKE DISTRICT (1:75,000)
BORDERS OF ENGLAND & SCOTLAND (1:200,000)
BURNS COUNTRY (1:200,000)
HEART OF SCOTLAND (1:200,000)
LOCH LOMOND & TROSSACHS (1:150,000)
PERTHSHIRE (1:150,000)
FORT WILLIAM, BEN NEVIS, GLEN COE (1:185,000)
IONA (1:10,000) & MULL (1:115,000)
GRAMPIAN HIGHLANDS (1:185,000)
LOCH NESS & INVERNESS (1:150,000)
AVIEMORE & SPEY VALLEY (1:150,000)
SKYE & LOCHALSH (1:130,000)
CAITHNESS & SUTHERLAND (1:185,000)
OUTER HEBRIDES (1:125,000)
ORKNEY & SHETLAND (1,128,000)
SCOTLAND (1:500,000)
GREAT BRITAIN (1:1,100,000)

ROAD ATLAS

MOTORING IN THE SOUTH (1:200,000)

EUROPEAN LEISURE MAPS

EUROPE (1:3,100,00)
BENELUX (1:600,000)
FRANCE (1:1,000,000)
GERMANY (1:1,000,000)
GREECE & THE AEGEAN (1:1,000,000)
IRELAND (1:625,000)
ITALY (1:1,000,000)
MEDITERRANEAN CRUISING (1:5,000,000)
SCANDINAVIA (1:2,600,000)
SPAIN & PORTUGAL (1:1,000,000)
THE ALPS (1:1,000,000)

ESTATE PUBLICATIONS are also
sole distributors in the U.K. for:
ORDNANCE SURVEY, Republic of Ireland
ORDNANCE SURVEY, Northern Ireland

Catalogue and prices from ESTATE PUBLICATIONS,
Bridewell House, Tenterden, Kent TN30 6JB.

CONTENT

One-way street	⟶	Post Office	●
Pedestrian Precinct	▨	Public Convenience	Ⓒ
Car Park	🅿	Place of Worship	✛

ADMINISTRATIVE DISTRICTS

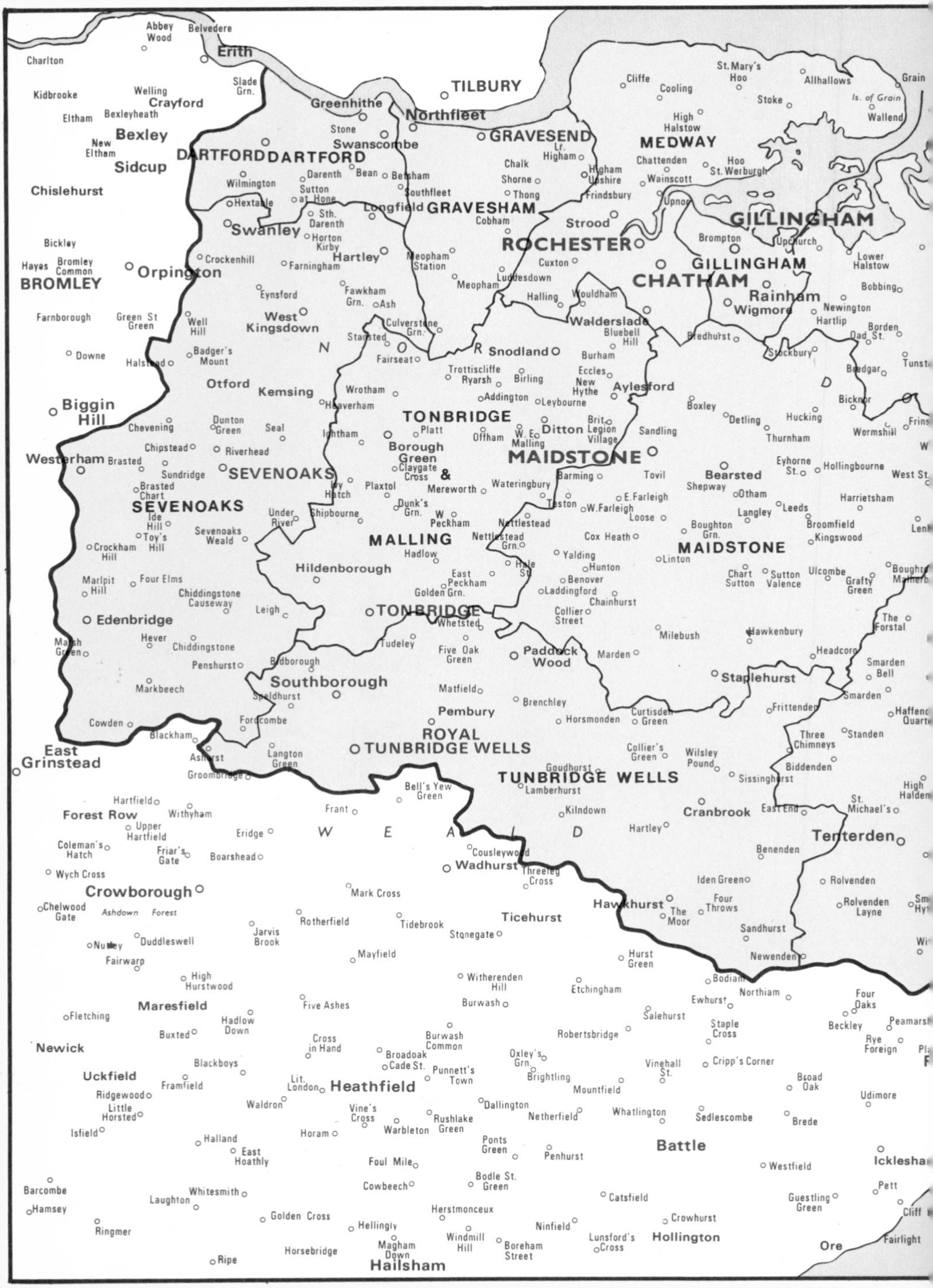
Abbey Wood
Belvedere
Erith
Charlton
Kidbrooke
Welling
Crayford
Slade Grn.
TILBURY
Greenhithe
Northfleet
Cliffe
Cooling
St. Mary's Hoo
Allhallows
Grain
Is. of Grain
Stoke
Wallend
Eltham
Bexleyheath
New Eltham
Bexley
Sidcup
DARTFORD
DARTFORD
Stone
Swanscombe
GRAVESEND
Lr. Higham
High Halstow
MEDWAY
Chattenden
Hoo St. Werburgh
Chalk
Shorne
Higham Upshire
Wainscott
Thong
Frindsbury
Upnor
Chislehurst
Wilmington
Darenth
Bean
Betsham
Southfleet
Sutton at Hone
Hextable
Longfield
GRAVESHAM
Cobham
Strood
GILLINGHAM
Sth. Darenth
Swanley
Horton Kirby
Hartley
Meopham Station
ROCHESTER
Brompton
GILLINGHAM
Upchurch
Lower Halstow
Bickley
Crockenhill
Farningham
Cuxton
Hayes
Bromley Common
Orpington
BROMLEY
Meopham
Luddesdown
CHATHAM
Rainham
Bobbing
Eynsford
Fawkham Grn.
Ash
Halling
Wouldham
Wigmore
Newington
Hartlip
Borden
Oad St.
Farnborough
Green St Green
Well Hill
West Kingsdown
Culverstone Grn.
Walderslade
Bluebell Hill
Bredhurst
Stockbury
Downe
Halstead
Badger's Mount
Stansted
Fairseat
Snodland
Burham
Bredgar
Tunst
Trottiscliffe
Ryarsh
Birling
Eccles
New Hythe
Aylesford
Otford
Kemsing
Wrotham
Addington
Leybourne
Bicknor
Biggin Hill
Heaverham
Boxley
Detling
Hucking
TONBRIDGE
Chevening
Dunton Green
Seal
Ightham
Platt
Offham
W. Malling
Ditton
Brit. Legion Village
Sandling
Thurnham
Wormshill
Chipstead
Riverhead
Borough Green
MAIDSTONE
Westerham
Brasted
Claygate Cross
&
Eyhorne St.
Hollingbourne
Sundridge
SEVENOAKS
Barming
Tovil
Bearsted
West St
Brasted Chart
Ivy Hatch
Plaxtol
Mereworth
Wateringbury
Shepway
SEVENOAKS
Teston
E. Farleigh
W. Farleigh
Otham
Harrietsham
Ide Hill
Dunk's Grn.
Under River
Shipbourne
W Peckham
Nettlestead
Loose
Langley
Leeds
Toy's Hill
Sevenoaks Weald
Boughton Grn.
Broomfield
Kingswood
Crockham Hill
MALLING
Nettlestead Grn.
Cox Heath
MAIDSTONE
Hadlow
Yalding
Linton
Hildenborough
Hale St.
Hunton
Ulcombe
Marlpit Hill
Four Elms
East Peckham
Benover
Chart Sutton
Sutton Valence
Grafty Green
Chiddingstone Causeway
Golden Grn.
Laddingford
Chainhurst
Edenbridge
Leigh
TONBRIDGE
Whetsted
Collier Street
The Forstal
Hever
Marsh Green
Chiddingstone
Tudeley
Five Oak Green
Paddock Wood
Marden
Milebush
Hawkenbury
Headcorn
Penshurst
Bidborough
Smarden Bell
Southborough
Staplehurst
Markbeech
Speldhurst
Matfield
Smarden
Brenchley
Frittenden
Pembury
Horsmonden
Curtisden Green
Cowden
Fordcombe
ROYAL TUNBRIDGE WELLS
Three Chimneys
Standen
East Grinstead
Blackham
Ashurst
Langton Green
Collier's Green
Wilsley Pound
Biddenden
Groombridge
Goudhurst
TUNBRIDGE WELLS
Sissinghurst
Hartfield
Bell's Yew Green
Lamberhurst
Forest Row
Withyham
Frant
Kilndown
Cranbrook
East End
St. Michael's
Upper Hartfield
Eridge
W E A L D
Hartley
Tenterden
Coleman's Hatch
Friar's Gate
Boarshead
Cousley Wood
Benenden
Wych Cross
Wadhurst
Threeleg Cross
Iden Green
Rolvenden
Crowborough
Mark Cross
Hawkhurst
The Moor
Four Throws
Rolvenden Layne
Chelwood Gate
Ashdown Forest
Rotherfield
Tidebrook
Ticehurst
Jarvis Brook
Stonegate
Sandhurst
Nutley
Duddleswell
Mayfield
Hurst Green
Newenden
Fairwarp
Witherenden Hill
Bodiam
High Hurstwood
Etchingham
Northiam
Four Oaks
Ewhurst
Fletching
Maresfield
Five Ashes
Burwash
Salehurst
Hadlow Down
Staple Cross
Beckley
Peasmarsh
Newick
Buxted
Cross in Hand
Burwash Common
Robertsbridge
Rye Foreign
Blackboys
Broadoak
Cade St.
Oxley's Grn.
Vinehall St.
Cripp's Corner
Uckfield
Framfield
Lit. London
Heathfield
Punnett's Town
Brightling
Mountfield
Broad Oak
Ridgewood
Little Horsted
Udimore
Waldron
Vine's Cross
Rushlake Green
Dallington
Netherfield
Whatlington
Sedlescombe
Brede
Isfield
Warbleton
Halland
Horam
Ponts Green
Battle
East Hoathly
Penhurst
Foul Mile
Westfield
Icklesham
Barcombe
Cowbeech
Bodle St. Green
Whitesmith
Pett
Laughton
Catsfield
Guestling Green
Hamsey
Golden Cross
Herstmonceux
Crowhurst
Cliff End
Hellingly
Ninfield
Ringmer
Windmill Hill
Lunsford's Cross
Hollington
Fairlight
Horsebridge
Magham Down
Boreham Street
Ore
Ripe
Hailsham

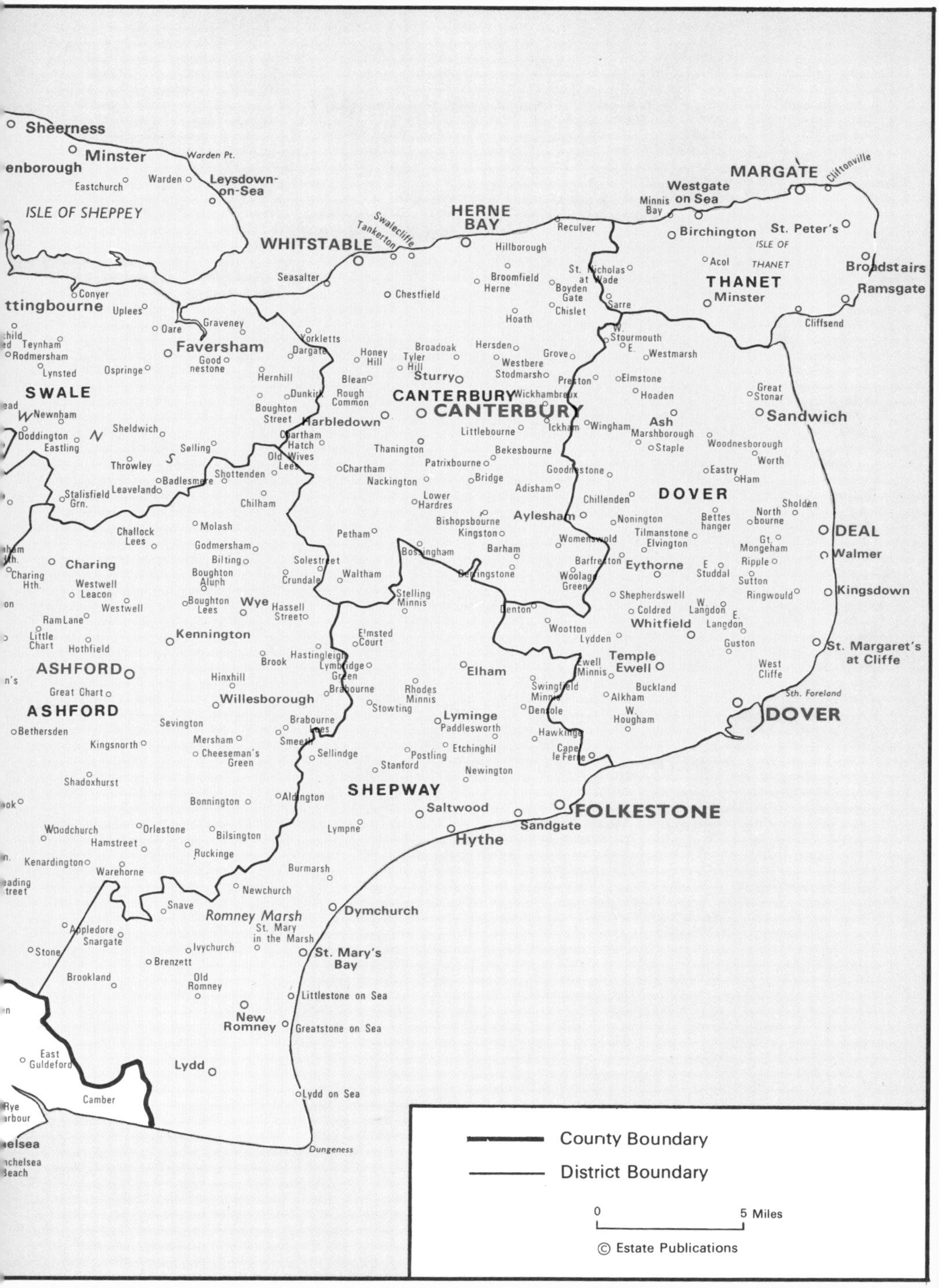

Sheerness
Minster
enborough
Warden Pt.
Eastchurch
Warden
Leysdown-on-Sea
ISLE OF SHEPPEY

MARGATE
Cliftonville
Westgate on Sea
Minnis Bay
HERNE BAY
Reculver
Birchington
St. Peter's
WHITSTABLE
Tankerton
Swalecliffe
Hillborough
ISLE OF THANET
Acol
Broadstairs
ttingbourne
Uplees
Seasalter
Chestfield
Broomfield
Herne
St. Nicholas at Wade
Boyden Gate
Sarre
THANET
Minster
Ramsgate
Cliffsend
child
Teynham
Rodmersham
Lynsted
Ospringe
Oare
Graveney
Yorkletts
Dargate
Hoath
Chislet
W.
E.
Stourmouth
Westmarsh
Faversham
Good
nestone
Hernhill
Honey
Hill
Broadoak
Tyler
Hill
Hersden
Grove
Westbere
Stodmarsh
Preston
Great Stonar
Sandwich
SWALE
Newnham
Boughton
Street
Dunkirk
Rough
Common
Blean
Sturry
Elmstone
Hoaden
Sheldwick
Harbledown
CANTERBURY
CANTERBURY
Ash
Marshborough
Woodnesborough
Worth
Doddington
Eastling
Throwley
Selling
Chartham
Hatch
Old Wives
Lees
Littlebourne
Ickham
Wingham
Staple
Eastry
Ham
Stalisfield
Grn.
Leaveland
Badlesmere
Shottenden
Chilham
Thanington
Chartham
Nackington
Patrixbourne
Bekesbourne
Bridge
Adisham
Goodnestone
Chillenden
Woodnesborough
North
bourne
Sholden
DEAL
Challock
Lees
Molash
Godmersham
Bilting
Petham
Lower
Hardres
Bishopsbourne
Kingston
Barham
Aylesham
Womenswold
Nonington
Tilmanstone
Elvington
Bettes
hanger
Gt.
Mongeham
Ripple
Walmer
Charing
Hth.
Charing
Westwell
Leacon
Westwell
Boughton
Aluph
Boughton
Lees
Wye
Hassell
Street
Crundale
Waltham
Solestreet
Bossingham
Kingstone
Denton
DOVER
Barfreston
Woolage
Green
Eythorne
E.
Studdal
W.
Sutton
Ringwould
Kingsdown
RamLane
Little
Chart
Hothfield
Kennington
Elmsted
Court
Stelling
Minnis
Wootton
Lydden
Shepherdswell
Coldred
Whitfield
Langdon
Guston
St. Margaret's
at Cliffe
ASHFORD
ASHFORD
Great Chart
Brook
Hastingleigh
Lymbridge
Green
Brabourne
Elham
Ewell
Minnis
Temple
Ewell
Buckland
West
Cliffe
Sth. Foreland
Bethersden
Kingsnorth
Hinxhill
Willesborough
Rhodes
Minnis
Stowting
Swingfield
Minnis
Denfole
Alkham
W.
Hougham
DOVER
Shadoxhurst
Sevington
Mersham
Cheeseman's
Green
Brabourne
Lees
Smeeth
Lyminge
Paddlesworth
Etchinghil
Hawkinge
Cape
le Ferne
ook
Woodchurch
Orlestone
Bilsington
Bonnington
Aldington
Sellindge
Stanford
Postling
Newington
SHEPWAY
Saltwood
FOLKESTONE
Hamstreet
Ruckinge
Lympne
Sandgate
Kenardington
Warehorne
Burmarsh
Hythe
reading
street
Snave
Newchurch
Romney Marsh
St. Mary
in the Marsh
Dymchurch
Appledore
Snargate
Ivychurch
St. Mary's
Bay
Stone
Brenzett
Old
Romney
Littlestone on Sea
Brookland
New
Romney
Greatstone on Sea
East
Guldeford
Lydd
Lydd on Sea
Camber
Rye
arbour
Dungeness
elsea
nchelsea
Beach

	County Boundary
	District Boundary

0   5 Miles

© Estate Publications

POSTCODE DISTRICTS

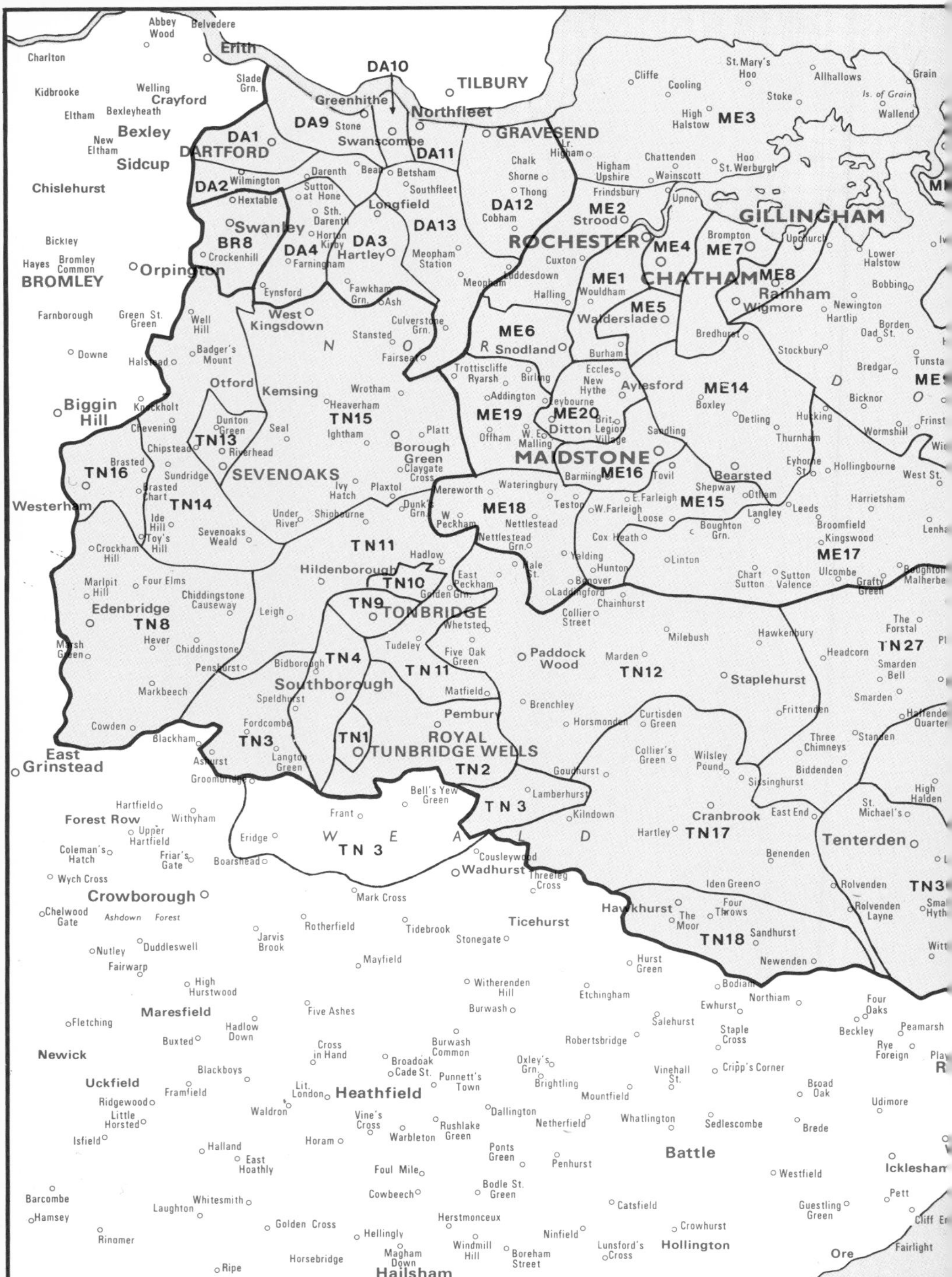
Abbey Wood
Belvedere
Erith
Charlton
Slade Grn.
DA10
TILBURY
Cliffe
Cooling
St. Mary's Hoo
Allhallows
Grain
Kidbrooke
Welling
Crayford
Greenhithe
Stoke
Is. of Grain
Eltham
Bexleyheath
DA9
Stone
Northfleet
High Halstow
ME3
Wallend
New Eltham
Bexley
DA1
Swanscombe
GRAVESEND
Lr. Higham
DARTFORD
DA11
Chalk
Chattenden
Hoo
St. Werburgh
Sidcup
Darenth
Bean
Betsham
Higham Upshire
Wainscott
Shorne
Chislehurst
DA2
Wilmington
Southfleet
Thong
Frindsbury
Sutton at Hone
Upnor
Hextable
Longfield
DA12
ME2
GILLINGHAM
Sth. Darenth
Cobham
Strood
Bickley
Swanley
DA13
BR8
Horton Kirby
DA3
ROCHESTER
ME4
ME7
Brompton
Upchurch
Hayes
Bromley Common
Crockenhill
DA4
Hartley
Meopham Station
Lower Halstow
Orpington
Farningham
Cuxton
BROMLEY
Luddesdown
ME1
CHATHAM
ME8
Eynsford
Fawkham Grn.
Ash
Meopham
Bobbing
Rainham
Halling
Wouldham
Farnborough
Green St. Green
West Kingsdown
Wigmore
Newington
ME5
Hartlip
Well Hill
Culverstone Grn.
Walderslade
Borden
Oad St.
Stansted
ME6
Bredhurst
Downe
Badger's Mount
N
O
R
Snodland
Stockbury
Tunstall
Halstead
Fairseat
Burham
Bredgar
ME9
Trottiscliffe
Eccles
Ryarsh
Birling
New Hythe
Otford
Wrotham
Aylesford
ME14
D
Kemsing
Addington
Bicknor
Knockholt
Heaverham
Larkfield
Boxley
Biggin Hill
Dunton Green
Seal
TN15
ME19
ME20
Detling
Hucking
Frinsted
Chevening
Platt
Ditton
Brit. Legion Village
Wormshill
Ightham
Offham
W. Malling
Sandling
Thurnham
TN13
Borough Green
Chipstead
Riverhead
MAIDSTONE
Wichling
Claygate Cross
Brasted
Eyhorne St.
Hollingbourne
TN16
Sundridge
SEVENOAKS
Barming
ME16
Tovil
Bearsted
West St.
Brasted Chart
Ivy Hatch
Plaxtol
Mereworth
Wateringbury
Shepway
Westerham
TN14
E. Farleigh
Otham
ME15
Harrietsham
Dunk's Grn.
ME18
Teston
Leeds
Ide Hill
Under River
Shipbourne
W. Peckham
Nettlestead
W. Farleigh
Langley
Broomfield
Toy's Hill
Sevenoaks Weald
Loose
Boughton Grn.
Kingswood
Lenham
Crockham Hill
TN11
Nettlestead Grn.
Cox Heath
ME17
Hadlow
Yalding
Linton
Marlpit Hill
Four Elms
Hildenborough
East Peckham
Hale St.
Hunton
Chart Sutton
Sutton Valence
Ulcombe
Grafty Green
Boughton Malherbe
Chiddingstone Causeway
TN10
Golden Grn.
Benover
Edenbridge
TN9
TONBRIDGE
Laddingford
Chainhurst
TN8
Leigh
Collier Street
The Forstal
Whetsted
Milebush
Hawkenbury
Marsh Green
Hever
Chiddingstone
Tudeley
Five Oak Green
Paddock Wood
Marden
Headcorn
TN27
Penshurst
Bidborough
TN4
TN11
TN12
Staplehurst
Smarden Bell
Southborough
Matfield
Markbeech
Speldhurst
Smarden
Brenchley
Frittenden
Pembury
Fordcombe
Curtisden Green
Cowden
Horsmonden
Haffenden Quarter
Blackham
TN1
ROYAL TUNBRIDGE WELLS
Three Chimneys
Standen
TN3
Collier's Green
Wilsley Pound
East Grinstead
Ashurst
Langton Green
TN2
Goudhurst
Biddenden
Groombridge
Bell's Yew Green
Lamberhurst
Sissinghurst
High Halden
Hartfield
TN 3
Frant
Kilndown
Cranbrook
East End
St. Michael's
Forest Row
Withyham
Upper Hartfield
W
E
A
L
D
Hartley
TN17
Tenterden
Eridge
TN 3
Benenden
Coleman's Hatch
Friar's Gate
Boarshead
Cousleywood
Wadhurst
Threeleg Cross
Iden Green
Rolvenden
TN30
Wych Cross
Crowborough
Mark Cross
Hawkhurst
Four Throws
Rolvenden Layne
Small Hythe
Chelwood Gate
Ashdown Forest
The Moor
Ticehurst
Rotherfield
Tidebrook
Jarvis Brook
Sandhurst
Stonegate
TN18
Nutley
Duddleswell
Hurst Green
Newenden
Wittersham
Fairwarp
Mayfield
High Hurstwood
Witherenden Hill
Bodiam
Etchingham
Northiam
Four Oaks
Burwash
Ewhurst
Maresfield
Five Ashes
Salehurst
Fletching
Hadlow Down
Staple Cross
Beckley
Peasmarsh
Buxted
Cross in Hand
Burwash Common
Robertsbridge
Rye Foreign
Newick
Broadoak
Oxley's Grn.
Blackboys
Cade St.
Vinehall St.
Cripp's Corner
Punnett's Town
Brightling
Uckfield
Lit. London
Heathfield
Broad Oak
Framfield
Mountfield
Ridgewood
Udimore
Waldron
Dallington
Little Horsted
Vine's Cross
Rushlake Green
Netherfield
Whatlington
Sedlescombe
Brede
Isfield
Warbleton
Horam
Halland
Ponts Green
Battle
East Hoathly
Penhurst
Icklesham
Foul Mile
Westfield
Bodle St. Green
Barcombe
Cowbeech
Pett
Whitesmith
Catsfield
Guestling Green
Laughton
Hamsey
Herstmonceux
Crowhurst
Cliff End
Golden Cross
Ringmer
Hellingly
Ninfield
Lunsford's Cross
Hollington
Windmill Hill
Fairlight
Magham Down
Boreham Street
Ore
Horsebridge
Ripe
Hailsham

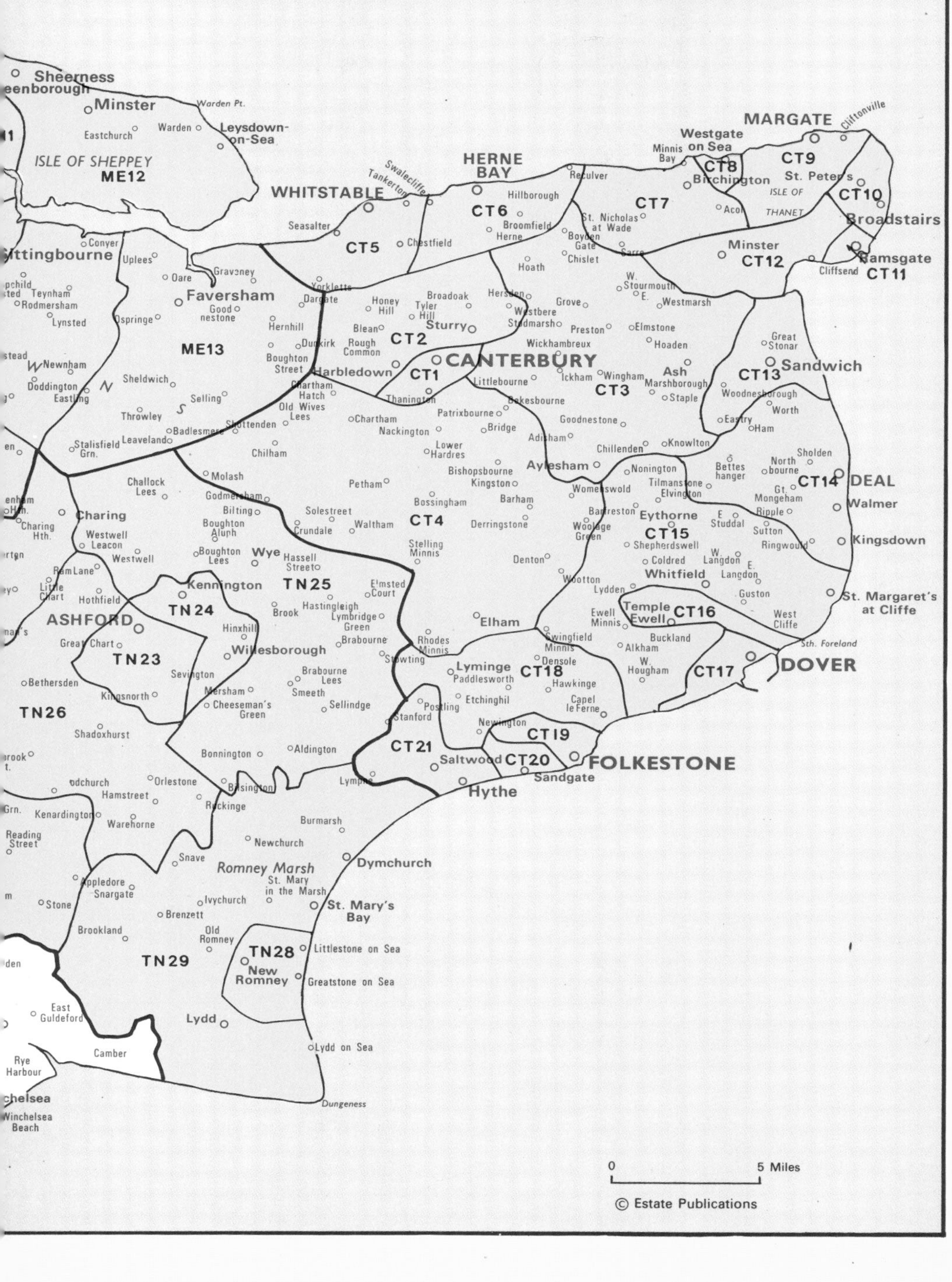

The Postcode District boundaries are reproduced by permission of the Post Office

Sheerness
eenborough
Minster
Eastchurch
Warden Pt.
Warden
Leysdown-
on-Sea

MARGATE
Westgate
Cliftonville
Minnis on Sea
Minnis
Bay
CT8
CT9
St. Peters
Birchington
ISLE OF
THANET
CT10
Broadstairs

ISLE OF SHEPPEY
ME12

WHITSTABLE
Swalecliffe
Tankerton
HERNE
BAY
Hillborough
Reculver
CT6
CT7
Acol
Minster
CT12
Ramsgate
Cliffsend
CT11

ttingbourne
Conyer
Uplees
Seasalter
Chestfield
CT5
Broomfield
St. Nicholas
at Wade
Boyden
Gate
Sarre
W.
Stourmouth
E.
Westmarsh

Graveney
Yorkletts
Dargate
Herne
Hoath
Chislet

Faversham
Good
nestone
Honey
Hill
Broadoak
Tyler
Hill
Herson
Grove
Elmstone
Hoaden
Great
Stonar
Sandwich

child
ted
Teynham
Rodmersham
Lynsted
Ospringe
Hernhill
Dunkirk
Blean
Sturry
Westbere
Stodmarsh
Preston
Wickhambreux
Woodnesborough
CT13

stead
Newnham
ME13
Boughton
Street
Rough
Common
Hartham
Hatch
Harbledown
CT2
CANTERBURY
Littlebourne
Ickham
Wingham
Ash
Marshborough
Staple
Worth
Ham
Eastry

Doddington
Eastling
Sheldwich
N
Old Wives
Lees
CT1
CT3
Bekesbourne
Goodnestone
Knowlton

en
Throwley
Stalisfield
Grn.
Leaveland
Badlesmere
Slottenden
Chilham
Chartham
Nackington
Lower
Hardres
Bridge
Patrixbourne
Adisham
Chillenden
North
bourne
Sholden
DEAL

enham
m
Challock
Lees
Molash
Godmersham
Petham
Bishopsbourne
Kingston
Aylesham
Nonington
Tilmanstone
Elvington
Bettes
hanger
Gt.
Mongeham
Ripple
CT14
Walmer

Charing
Hth.
Charing
Westwell
Leacon
Bilting
Boughton
Aluph
Solestreet
Waltham
Bossingham
Barham
Womenswold
Barfreston
Woolage
Green
Eythorne
CT15
E
Studdal
Sutton
Ringwould
Kingsdown

PumLane
Little
Chart
Westwell
Boughton
Lees
Wye
Hassell
Street
CT4
Derringstone
Shepherdswell
Coldred
W.
Langdon
E.
Langdon
St. Margaret's
at Cliffe

ASHFORD
Hothfield
Kennington
TN25
Stelling
Minnis
Denton
Wootton
Lydden
Whitfield
Guston
West
Cliffe

Great Chart
TN24
Brook
Hastingleigh
Lymbridge
Green
Brabourne
Elmsted
Court
Elham
Ewell
Minnis
Temple
Ewell
CT16
Sth. Foreland

mar's
TN23
Hinxhill
Willesborough
Rhodes
Minnis
Wingfield
Minnis
Buckland
Alkham
W.
Hougham
CT17
DOVER

Bethersden
Sevington
Brabourne
Lees
Smeeth
Stowting
Lyminge
Paddlesworth
CT18
Densole
Hawkinge
Capel
le Ferne
FOLKESTONE

TN26
Kingsnorth
Mersham
Cheeseman's
Green
Sellindge
Postling
Stanford
Etchinghil
Newington
Sandgate

Shadoxhurst
Bonnington
Aldington
CT21
Saltwood
CT20
CT19

arook
t.
odchurch
Orlestone
Hamstreet
Bilsington
Ruckinge
Lymp
Hythe

Grn.
Kenardington
Warehorne
Burmarsh

Reading
Street
Snave
Newchurch
Dymchurch

Appledore
Snargate
Ivychurch
Brenzett
Old
Romney
Romney Marsh
St. Mary
in the Marsh
St. Mary's
Bay

m
Stone
Brookland
TN28
New
Romney
Littlestone on Sea
Greatstone on Sea

lden
East
Guldeford
TN29
Lydd
Camber
Lydd on Sea

Rye
Harbour
chelsea
Winchelsea
Beach
Dungeness

0   5 Miles

© Estate Publications

POPULATIONS OF DISTRICTS

County of Kent, population 1,467,619	Maidstone 130,496
Ashford 85,968	Rochester Upon Medway 143,846
Canterbury 117,169	Sevenoaks 109,871
Dartford 78,345	Shepway 86,503
Dover 100,987	Swale 109,647
Gillingham 93,734	Thanet ¹21,720
Gravesham 95,976	Tonbridge and Malling 96,848
	Tunbridge Wells 96,509

Acol 193   13 D1
Addington 676   10 B2
Adisham 583   13 C2
Aldington 1,133   12 B3
Alkham 575   13 C3
Allhallows 1,878   11 D1
Appledore 661   15 C3
Ash (nr Dartford) 5,569   10 B2
Ash (nr Canterbury) 2,636   13 D2
Ashford, E.C. Wed 39,994   12 A3
Ashurst   10 A3
Aylesford 8,096   11 C2
Aylesham 3,899   13 C2

Badger's Mount   10 A2
Badlesmere 126   12 B2
Bapchild 1,008   12 A1
Barfreston   13 C2
Barham 1,239   13 C2
Barming 2,040   11 C2
Bean   10 B1
Bearsted 5,811   11 C2
Bekesbourne 580   13 C2
Beltring   14 A1
Benenden 1,698   14 B2
Benover   14 B1
Bethersden 1,273   12 A3
Betsham   10 B1
Betteshanger   13 D2
Bicknor 74   11 D2
Bidborough 939   10 B3
Biddenden 2,229   15 C2
Bilsington 232   15 D2
Bilting   12 B2
Birchington   13 C1
Birling 1,872   11 C2
Bishopsbourne 234   13 C2
Blean 1,859   12 B1
Bluebell Hill   11 C2
Bobbing 1,949   11 D2
Bonnington 101   12 B3
Borden 1,759   11 D2
Borough Green 3,213   10 B2
Bossingham   13 C2
Boughton Aluph 1,053   12 B2
Boughton Grn   11 C3
Boughton Lees   12 B2
Boughton Malherbe 344   12 A2
Boughton Monchesea 1,678   *
Boughton Street 1,827   12 B2
Boxley 3,097   11 C2
Boyden Gate   13 C1
Brabourne 1,123   12 B3
Brabourne Lees   12 B3
Brasted 1,313   10 A2
Brasted Chart   10 A2
Bredgar 654   11 D2
Bredhurst 273   11 C2
Brenchley 2,582   14 A2
Brenzett 316   15 D3
Bridge 1,302   13 C2
British Legion Village   11 C2
Broadoak   13 C1
Broadstairs, E.C. Wed 23,447   13 D1
Broad Street   11 D2
Bromley   10 A1
Brompton   11 C1
Brook 335   12 B3
Brookland 492   15 D3
Broomfield (nr Herne Bay)   13 C1
Broomfield (nr Maidstone) 1,657   15 C1
Buckland 78   13 D2
Burham 1,721   11 C2
Burmarsh 316   15 D3

Canterbury, E.C. Thurs 35,429   12 B2
Capel le Ferne 1,760   13 C3
Chainhurst   14 B1
Chalk   11 C1
Challock Lees 743   12 A2
Charing, E.C. Wed 2,661   12 A2
Charing Hth   12 A2
Chartham 3,646   12 B2
Chartham Hatch   12 B2
Chart Sutton 873   14 B1
Chatham, E.C. Wed 43,915   11 C1
Chattenden   11 C1
Cheeseman's Green   12 B3
Cheriton   13 C3
Chestfield   12 B1

Chevening 3,028   10 A2
Chiddingstone 812   10 A3
Chiddingstone Causeway   10 A3
Chilham 1,527   12 B2
Chillenden   13 C2
Chipstead   10 A2
Chislet 670   13 C1
Cliffe 4,748   11 C1
Cliffe Woods   11 C1
Cliffsend   13 D1
Cliftonville   13 D1
Cobham 1,340   11 C1
Coldred   13 C3
Collier's Grn   14 B2
Collier St   14 B2
Conyer   12 A1
Cooling 160   11 C1
Cowden 785   10 A3
Coxheath 4,230   11 C3
Cranbrook 5,344   14 B2
Cranbrook Common   14 B2
Crockenhill   10 A1
Crockham Hill   10 A3
Cross-at-Hand   14 B1
Crundale 201   12 B2
Culverstone Grn   10 B2
Curtisden Green   14 B2
Cuxton 3,011   11 C1

Darenth 4,942   10 B1
Dargate   12 B1
Dartford, E.C. Wed 44,003   10 A1
Deal, E.C. Thurs 26,109   13 D2
Densole   13 C3
Denton 285   13 C2
Derringstone   13 C2
Detling 800   11 C2
Ditton 5,168   11 C2
Doddington 447   12 A2
Dover, E.C. Wed 32,830   13 D3
Dumpton   13 D1
Dunkirk 1,015   12 B2
Dunk's Grn   10 B2
Dunton Green 1,738   10 A2
Dymchurch 3,384   15 D3

Easole Street 13 C2
Eastchurch   12 A1
East End   14 B2
East Farleigh 1,250   11 C2
East Langdon   13 D3
Eastling 353   12 A2
East Malling (incl Larkfield) 11,573   11 C2
East Peckham 3,287   14 A1
Eastry 2,478   13 D2
East Stourmouth   13 C1
East Studdal   13 D2
Eastwell 129   *
Eccles   11 C2
Edenbridge 7,903   10 A3
Egerton 940   12 A2
Egerton Forstal   15 C1
Elham 1,321   13 C3
Elmsted Court 247   12 B3
Elmstone   13 C1
Elvington   13 D2
Etchinghill   13 C3
Ewell Minnis   13 C3
Eyhorne St   11 D2
Eynsford 3,050   10 B2
Eythorne 2,338   13 D2

Fairseat   10 B2
Farningham 1,319   10 B1
Faversham, E.C. Thurs 16,115   12 B1
Fawkham Grn 539   10 B1
Five Oak Green   14 A2
Folkestone, E.C. Wed 44,998   13 C3
Fordcombe   10 A3
Fordwich 177   13 C2
Four Elms   10 A3
Four Throws   14 B2
Frindsbury 4,983   11 C1
Frinsted 132   12 A2
Frittenden 819   14 B2

Gillingham, E.C. Wed 93,734   11 C1
Godmersham 341   12 B2
Golden Grn   14 A1

Goodnestone (nr Faversham) 51   *
Goodnestone (nr Canterbury) 351   13 C2
Goudhurst 2,673   14 B2
Grafty Green   15 C1
Grain   11 D1
Graveney 291   12 B1
Gravesend, E.C. Wed 52,905   10 B1
Great Chart 929   12 A3
Great Mongeham   13 D2
Great Stonar   13 D1
Greatstone-on-Sea   15 D3
Greenhill   13 C1
Greenhithe   10 B1
Grove   13 C1
Guston 1,134   13 D3

Hadlow 3,365   14 A1
Haffenden Quarter   15 C2
Hale St   14 A1
Halfway   12 A1
Halling 1,800   11 C2
Halstead 1,661   10 A2
Ham   13 D2
Hamstreet   15 D2
Harbledown 2,373   12 B2
Harrietsham 1,404   15 C1
Hartley (nr Cranbrook)   14 B2
Hartley (nr Dartford) 6,963   10 B1
Hartlip 655   11 D2
Hassell St   12 B3
Hastingleigh 234   12 B3
Hawkenbury   14 B2
Hawkhurst 3,968   14 B2
Hawkinge 1,954   13 C3
Hawley   10 B1
Headcorn 3,208   15 C2
Heaverham   10 B2
Herne   13 C1
Herne Bay, E.C. Thurs 28,143   13 C1
Hernhill 569   12 B1
Hersden   13 C1
Hever 1,033   10 A3
Hextable   10 A1
Higham 4,000   11 C1
High Brooms   14 A2
High Halden 1,241   15 C2
High Halstow 1,331   11 C1
Highsted   12 A1
Hildenborough 4,983   10 B3
Hillborough   13 C1
Hinxhill 46   12 B3
Hoaden   13 C2
Hoath 439   13 C1
Hollingbourne 965   11 D2
Honey Hill   12 B1
Hoo St, Werburgh 8,105   11 C1
Horsmonden 1,907   14 B2
Horton Kirby 2,938   10 B1
Hothfield 792   12 A3
Hucking 54   11 D2
Hunton 539   14 B1
Hythe, E.C. Wed 12,823   13 C3

Ickham 426   13 C2
Ide Hill   10 A3
Iden Green   14 B2
Ightham 1,731   10 B2
Isle of Grain 1,857   11 D1
Istead Rise   10 B1
Ivychurch 211   15 D3
Ivy Hatch   10 B2
Iwade 1,010   12 A1

Kemsing 4,127   10 B2
Kemsley   12 A1
Kenardington 250   15 C2
Kennington   12 B3
Kilndown (with Ringwould)   14 B2
Kingsdown   13 D2
Kingsgate   13 D1
Kingsnorth 6,097   12 A3
Kingston 518   13 C2
Kingswood   15 C1
Knockholt 1,238   10 A2
Knockholt Pound   10 A2
Knowlton   13 D2

Laddingford   14 B1

Place	Pop.	Ref
Lamberhurst **1,323**		14 B2
Langley **1,177**		11 D3
Langton Green		10 B3
Leaveland **107**		12 A2
Leeds **758**		11 D3
Leigh **1,446**		10 B3
Lenham **3,549**		12 A2
Lenham Hth		12 A2
Leybourne **1,670**		11 C2
Leysdown-on-Sea		12 B1
Linton **545**		14 B1
Littlebourne **1,452**		13 C2
Little Chart **234**		12 A3
Littlestone on Sea		15 D3
Longfield **2,218**		10 B1
Loose **2,459**		11 C3
Lower Halstow **1,048**		11 D1
Lower Hardres **420**		13 C2
Lower Higham		11 C1
Luddesdown **230**		11 C1
Lydd, E.C. Wed **4,729**		15 D3
Lydden **663**		13 C3
Lydd on Sea		15 D3
Lymbridge Grn		12 B3
Lyminge **2,573**		13 C3
Lympne **1,016**		12 B3
Lynsted **981**		12 A1
Maidstone, E.C. Wed **72,494**		11 C2
Maltman's Hill		12 A3
Manston		13 D1
Marden **3,294**		14 B2
Margate, E.C. Thurs **55,492**		13 D1
Markbeech		10 A3
Marlpit Hill		10 A3
Marsh Green		10 A3
Matfield		14 A2
Meopham **8,551**		10 B1
Meopham Station		10 B1
Mereworth **1,323**		10 B2
Mersham **1,070**		12 B3
Milebush		14 B2
Milstead **182**		12 A2
Milton Regis		12 A1
Minnis Bay		13 D1
Minster in Sheppey		12 A1
Minster (Thanet) **3,158**		13 D1
Molash **254**		12 B2
Monkton **540**		13 D1
Nackington		13 C2
Nettlestead **838**		11 C2
Nettlestead Grn		14 A1
New Ash Green		10 B1
Newchurch **301**		15 D3
Newenden **188**		14 B2
New Hythe		11 C2
Newingreen		12 B3
Newington (nr Hythe) **345**		13 C3
Newington (nr Rainham) **2,538**		11 D2
Newnham **347**		12 A2
New Romney, E.C. Wed **4,547**		15 D3
Nonington **653**		13 C2
Northbourne **674**		13 D2
Northfleet, E.C. Wed **26,310**		10 B1
Oad St		11 D2
Oare **526**		12 A1
Offham **803**		10 B2
Old Romney **205**		15 D3
Old Wives Lees		12 B2
Orlestone **1,125**		*
Orpington		10 A1
Ospringe **474**		12 A1
Otford **3,557**		10 A2
Otham **941**		11 C2
Paddlesworth **39**		13 C3
Paddock Wood **6,490**		14 A2
Patrixbourne **342**		13 C2
Pembury **6,285**		14 A2
Penshurst **1,749**		10 A3
Petham **675**		12 B2
Platt **1,270**		10 B2
Plaxtol **924**		10 B2
Pluckley **875**		12 A3
Pluckley Thorne		12 A3
Postling **176**		12 B3

Place	Pop.	Ref
Preston		12 B1
Preston **637**		13 C1
Queenborough (with Sheerness)		11 D1
Rainham		11 D1
Ramsgate, E.C. Thurs **43,397**		13 D1
Reading St		15 C3
Reculver		13 C1
Redbrook St		15 C2
Rhodes Minnis		13 C3
Richborough		13 D1
Ringwould **1,830**		13 D2
Ripple **310**		13 D2
Riverhead **1,765**		10 A2
Rochester, E.C. Wed **67,651**		11 C1
Rodmersham **362**		12 A1
Rolvenden **1,270**		15 C3
Rolvenden Layne		15 C3
Rough Common		12 B2
Ruckinge **704**		15 D2
Rusthall		10 B3
Ryarsh **696**		10 B2
St. Margaret's-at-Cliffe **2,312**		13 D3
St. Margarets' Bay		13 D3
St. Mary-in-the-Marsh **2,767**		15 D3
St. Mary's Bay		15 D3
St. Mary's Hoo **163**		11 C1
St. Michael's		15 C2
St. Nicholas-at-Wade **827**		13 C1
St. Peter's		13 D1
Saltwood **836**		13 C3
Sandgate		13 C3
Sandhurst **1,272**		14 B2
Sandling		11 C2
Sandwich, E.C. Thurs **4,254**		13 D2
Sarre **117**		13 C1
Seal **2,423**		10 B2
Seasalter		12 B1
Sellindge **1,243**		12 B3
Selling **674**		12 B2
Sevenoaks, E.C. Wed **17,241**		10 B2
Sevenoaks Weald **1,359**		10 A3
Sevington **176**		12 B3
Shadoxhurst **1,063**		12 A3
Sheerness (with Queenborough) **33,411**		12 A1
Sheldwich **432**		12 B2
Shepherdswell **1,744**		13 C2
Shepway		11 C2
Shipbourne **437**		10 B3
Sholden **739**		13 D2
Shoreham **2,003**		10 A2
Shorne **2,565**		11 C1
Shottenden		12 B2
Singlewell		10 B1
Sinkhurst Green		14 B2
Sissinghurst		14 B2
Sittingbourne, E.C. Wed **42,691**		12 A1
Small Hythe		15 C3
Smarden **1,156**		15 C2
Smarden Bell		15 C2
Smeeth **1,009**		12 B3
Snargate **111**		15 D3
Snave		15 D3
Snodland **5,836**		11 C2
Sole Street		12 B2
Southborough, E.C. Wed **10,069**		10 B3
Sth Darenth		10 B1
Southfleet **3,069**		10 B1
Speldhurst **4,213**		10 B3
Stalisfield Grn **198**		12 A2
Standen		15 C2
Stanford **476**		12 B3
Stansted **476**		10 B2
Staple **403**		13 C2
Staplehurst **5,818**		14 B2
Stelling Minnis **480**		13 C2
Stockbury **760**		11 D2
Stodmarsh		13 C1
Stoke **1,032**		11 D1
Stone (in Oxney) **373**		15 C3
Stone (nr Dartford) **8,560**		10 B1
Stourmouth (E.&W.) **235**		13 C2
Stowting **178**		12 B3
Strood		11 C1
Sturry **5,967**		13 C1

Place	Pop.	Ref
Sundridge **1,907**		10 A2
Sutton **858**		13 D2
Sutton-at-Hone **3,956**		10 B1
Sutton Valence **1,526**		14 B1
Swalecliffe		12 B1
Swanley **20,945**		10 A1
Swanscombe, E.C. Wed **8,876**		10 B1
Swingfield Minnis **1,177**		13 C3
Tankerton		12 B1
Temple Ewell **1,521**		13 C3
Tenterden, E.C. Wed **6,238**		15 C2
Teston **599**		11 C2
Teynham **3,211**		12 A1
Thanington **1,018**		13 C2
The Moor		14 B2
Thong		11 C1
Three Chimneys		14 B2
Throwley **293**		12 A2
Thurnham **2,407**		11 D2
Tilmanstone **375**		13 D2
Tonbridge, E.C. Wed **30,530**		14 A1
Tovil		11 C2
Toy's Hill		10 A3
Trottiscliffe **527**		10 B2
Tudeley		14 A2
Tunbridge Wells, E.C. Wed **44,992**		14 A2
Tunstall **734**		11 D2
Tyler Hill		13 C1
Ulcombe **836**		15 C1
Under River		10 B3
Upchurch **1,881**		11 D1
Uplees		12 A1
Upnor		11 C1
Upstreet		13 C1
Vigo Village		10 B2
Wainscott		11 C1
Walderslade		11 C2
Wallend		11 D1
Walmer		13 D2
Waltham **375**		12 B2
Warden		12 B1
Warehorne **300**		15 D3
Warren St		12 A2
Wateringbury **1,519**		11 C2
Well Hill		10 A2
Westbere **775**		13 C1
West Cliffe		13 D3
Westerham **4,333**		10 A2
West Farleigh **462**		11 C2
Westgate-on-Sea		13 D1
West Hougham		13 C3
West Kingsdown **4,909**		10 B2
West Langdon		13 D2
West Malling **2,371**		11 C2
Westmarsh		13 D1
West Peckham **293**		10 B2
West Stourmouth		13 C1
West St		12 A2
Westwell **949**		12 A2
Westwell Leacon		12 A2
Whetsted		14 A1
Whitfield **5,131**		13 D3
Whitstable, E.C. Wed **27,287**		12 B1
Wichling **102**		12 A2
Wickhambreux **459**		13 C2
Wigmore		11 C2
Willesborough		12 B3
Wilmington **6,938**		10 B1
Wilsley Pound		14 B2
Wingham **1,435**		13 C2
Wittersham **1,057**		15 C3
Womenswold **314**		13 C2
Woodchurch **1,662**		15 C2
Woodnesborough **967**		13 D2
Woolage Grn		13 C2
Wootton		13 C3
Wormshill **196**		11 D2
Worth **807**		13 D2
Wouldham **819**		11 C2
Wrotham **1,669**		10 B2
Wrotham Heath		10 B2
Wye **2,000**		12 B3
Yalding **2,767**		11 C3
Yorkletts		12 B1

Population figures in bold type   E.C. — Early Closing   * Not shown on map

Population figures are based upon the 1981 census and relate to the local authority ward or parish, as constituted at that date. Places with no population figure form part of a larger local authority area.

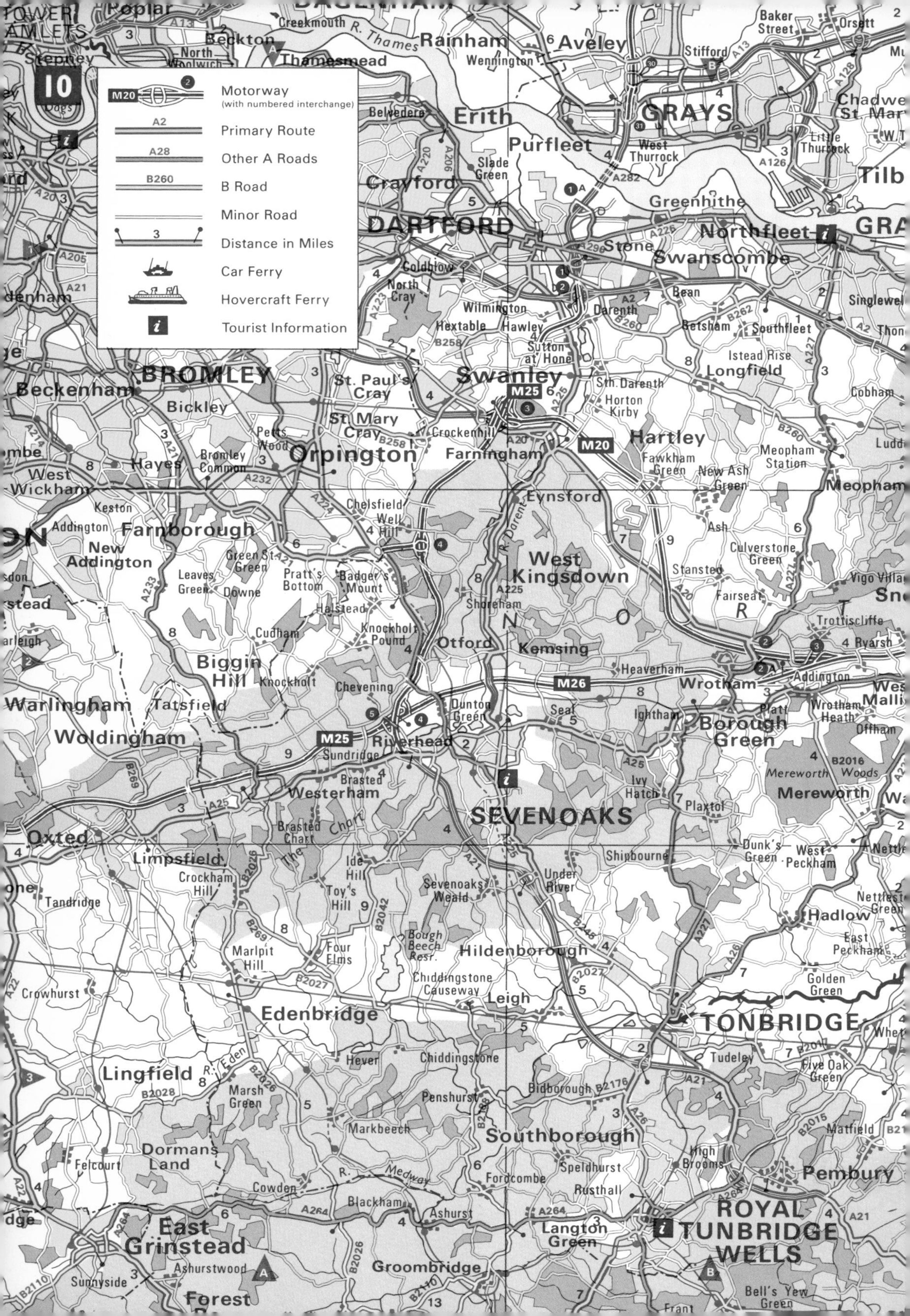

10
Motorway (with numbered interchange)
Primary Route
Other A Roads
B Road
Minor Road
Distance in Miles
Car Ferry
Hovercraft Ferry
Tourist Information
Poplar
Beckton
North Woolwich
Stepney
Creekmouth
R. Thames
Thamesmead
Rainham
Wennington
Aveley
Baker Street
Orsett
Stifford
GRAYS
West Thurrock
Little Thurrock
Chadwell St. Mary
Tilbury
Belvedere
Erith
Purfleet
Slade Green
Crayford
DARTFORD
Greenhithe
Northfleet
Stone
Swanscombe
Coldblow
North Cray
Wilmington
Hextable
Hawley
Sutton at Hone
Darenth
Bean
Betsham
Southfleet
Singlewell
Istead Rise
Longfield
Cobham
BROMLEY
Beckenham
Bickley
St. Paul's Cray
St. Mary Cray
Swanley
Sth. Darenth
Horton Kirby
Crockenhill
Farningham
Hartley
Fawkham Green
New Ash Green
Meopham Station
Meopham
Petts Wood
Orpington
Hayes
Bromley Common
West Wickham
Keston
Addington
New Addington
Farnborough
Chelsfield
Well Hill
Eynsford
West Kingsdown
Ash
Culverstone Green
Stansted
Fairseat
Vigo Village
Green St. Green
Leaves Green
Downe
Pratt's Bottom
Badger's Mount
Halstead
Shoreham
Knockholt Pound
Cudham
Biggin Hill
Knockholt
Otford
Kemsing
Heaverham
Trottiscliffe
Ryarsh
Addington
Wrotham
Wrotham Heath
West Malling
Offham
Chevening
Dunton Green
Seal
Ightham
Platt
Borough Green
Warlingham
Tatsfield
Woldingham
Riverhead
Sundridge
Brasted
Westerham
SEVENOAKS
Ivy Hatch
Plaxtol
Mereworth Woods
Mereworth
Oxted
Limpsfield
Brasted Chart
The Chart
Ide Hill
Toy's Hill
Sevenoaks Weald
Under River
Shipbourne
Dunk's Green
West Peckham
Crockham Hill
Tandridge
Hadlow
Nettlestead Green
East Peckham
Marlpit Hill
Four Elms
Bough Beech Resr.
Hildenborough
Chiddingstone Causeway
Leigh
Golden Green
Crowhurst
Edenbridge
TONBRIDGE
Lingfield
R. Eden
Hever
Chiddingstone
Tudeley
Five Oak Green
Marsh Green
Penshurst
Bidborough
Southborough
Markbeech
Matfield
Dormans Land
Felcourt
R. Medway
Cowden
Fordcombe
Speldhurst
High Brooms
Pembury
Rusthall
Blackham
Ashurst
Langton Green
ROYAL TUNBRIDGE WELLS
East Grinstead
Ashurstwood
Groombridge
Sunnyside
Forest Row
Bell's Yew Green
Frant

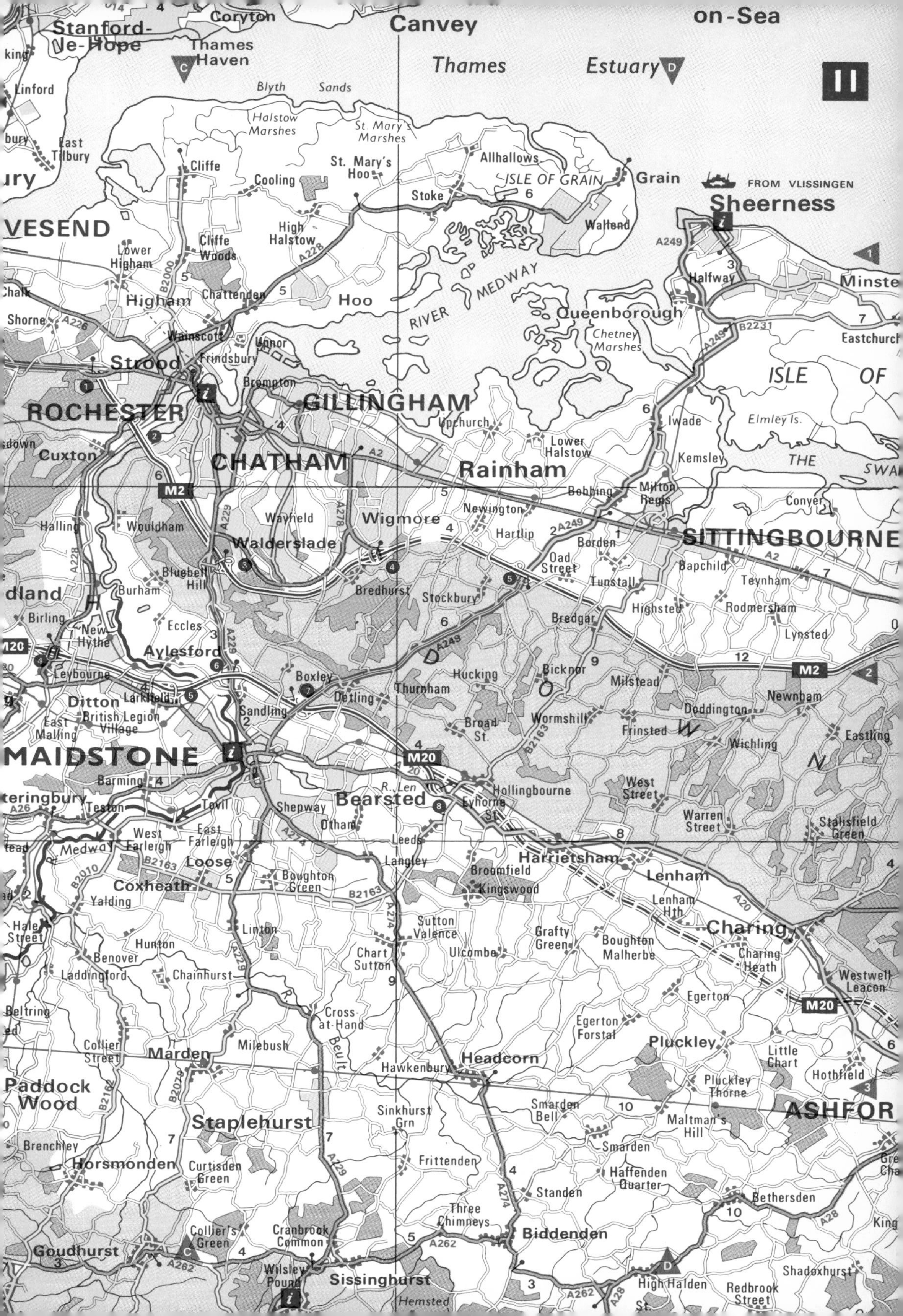
Stanford-le-Hope
Coryton
Canvey
on-Sea
Thames Haven
Thames
Estuary
Linford
Blyth Sands
Halstow Marshes
St. Mary's Marshes
East Tilbury
Cliffe
Cooling
St. Mary's Hoo
Allhallows
Grain
FROM VLISSINGEN
Sheerness
ISLE OF GRAIN
Stoke
Wallend
VESEND
High Halstow
Lower Higham
Cliffe Woods
A228
A249
Halfway
Minste
Higham
Chattenden
Hoo
RIVER MEDWAY
Queenborough
Shorne
A226
B2000
Chetney Marshes
B2231
Eastchurch
Wainscott
Upnor
Strood
Frindsbury
ISLE OF
Brompton
ROCHESTER
GILLINGHAM
Upchurch
Iwade
Elmley Is.
Cuxton
CHATHAM
A2
Lower Halstow
Kemsley
THE
SWA
Rainham
Bobbing
Milton Regis
Conyer
M2
Halling
Wouldham
Wayfield
A278
Wigmore
Newington
Hartlip
Borden
SITTINGBOURNE
A229
Walderslade
Oad Street
Bapchild
A2
Teynham
Bluebell Hill
Bredhurst
Stockbury
Tunstall
Burham
Highsted
Rodmersham
Birling
New Hythe
Eccles
Bredgar
Lynsted
M20
Aylesford
A249
Hucking
Bicknor
Milstead
M2
Leybourne
Boxley
Thurnham
Newnham
Ditton
Larkfield
Detling
Doddington
East Malling
British Legion Village
Sandling
Broad St.
Wormshill
Frinsted
Wichling
Eastling
MAIDSTONE
M20
B2163
DOWN
Barming
A20
R. Len
Hollingbourne
West Street
teringbury
Teston
Tovil
Bearsted
Eyhorne St.
Warren Street
Stalisfield Green
A26
Shepway
Otham
Medway
West Farleigh
East Farleigh
A274
Leeds
Harrietsham
Langley
Broomfield
Lenham
B2163
Loose
Boughton Green
Kingswood
B2010
Coxheath
Lenham Hth.
A20
Yalding
Charing
Sutton Valence
Grafty Green
Boughton Malherbe
Hale Street
Linton
Charing Heath
Hunton
Chart Sutton
Ulcombe
Benover
Westwell Leacon
Laddingford
Chainhurst
Egerton
Beltring
M20
Cross-at-Hand
Egerton Forstal
Collier Street
Milebush
Pluckley
Little Chart
Marden
Headcorn
R. Beult
Hawkenbury
Hothfield
Pluckley Thorne
Paddock Wood
B2162
B2079
Smarden Bell
Sinkhurst Grn
ASHFOR
Maltman's Hill
Staplehurst
A229
Brenchley
Smarden
Horsmonden
Curtisden Green
Frittenden
Haffenden Quarter
A274
Standen
Bethersden
Three Chimneys
Collier's Green
Cranbrook Common
Biddenden
A28
Goudhurst
A262
A262
Wilsley Pound
Sissinghurst
High Halden
Shadoxhurst
Redbrook Street
A262
A28
Hemsted

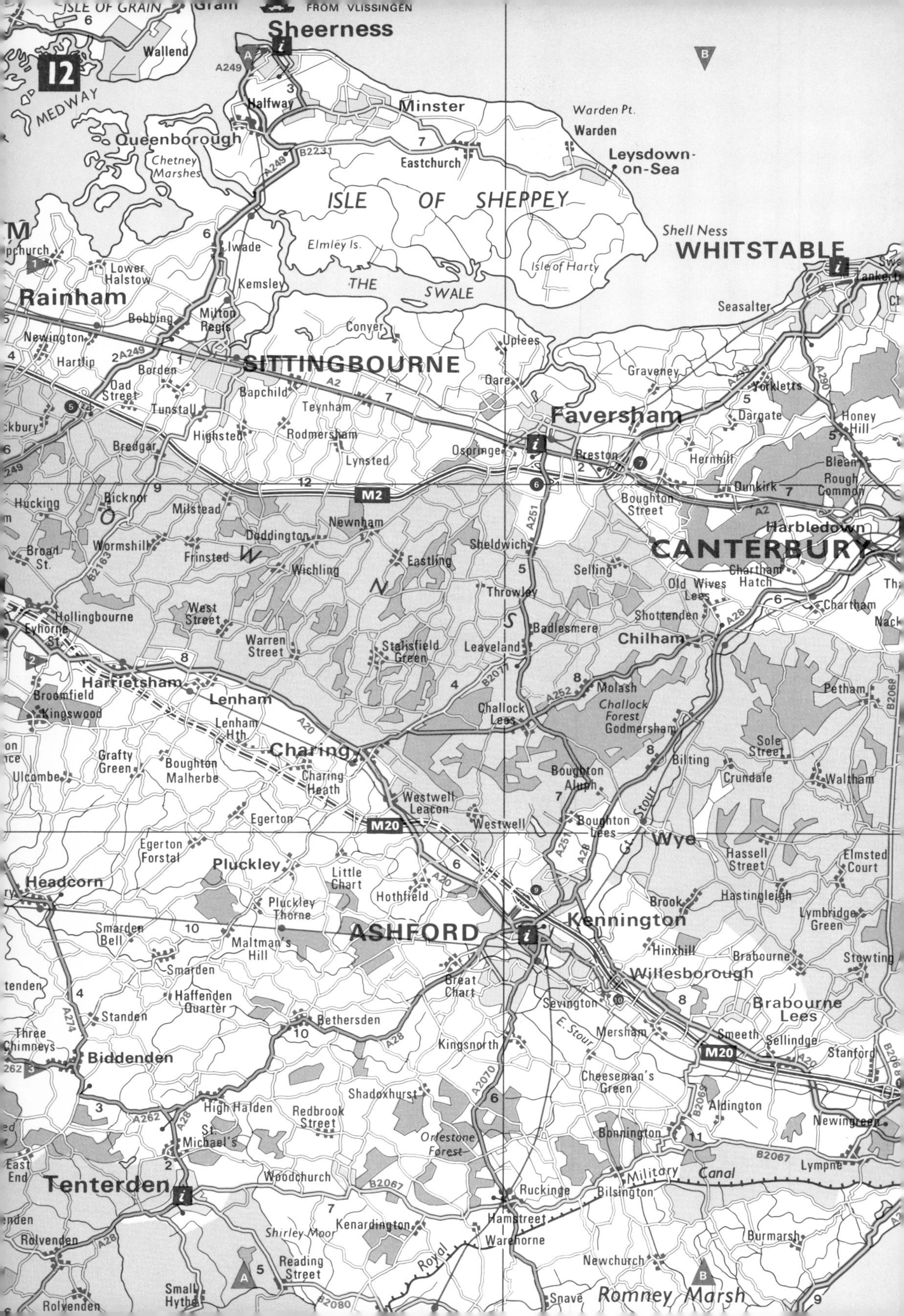

12
ISLE OF GRAIN
Grain
FROM VLISSINGEN
Sheerness
Wallend
MEDWAY
Halfway
Minster
Warden Pt.
Warden
Leysdown-on-Sea
Queenborough
Chetney Marshes
Eastchurch
ISLE OF SHEPPEY
Elmley Is.
Shell Ness
WHITSTABLE
Isle of Harty
Iwade
Lower Halstow
Kemsley
THE SWALE
Rainham
Bobbing
Milton Regis
Conyer
Seasalter
Newington
Uplees
Hartlip
Borden
SITTINGBOURNE
Graveney
Yorkletts
Oad Street
Bapchild
Oare
Teynham
Dargate
Tunstall
Faversham
Honey Hill
Highsted
Rodmersham
Bredgar
Ospringe
Preston
Hernhill
Lynsted
Blean
Rough Common
Hucking
Bicknor
Dunkirk
Milstead
Boughton Street
Newnham
Harbledown
Doddington
Broad St.
Wormshill
Sheldwich
CANTERBURY
Frinsted
Eastling
Wichling
Selling
Chartham Hatch
Old Wives Lees
Throwley
Chartham
Hollingbourne
West Street
Shottenden
Eyhorne St.
Badlesmere
Warren Street
Stalisfield Green
Leaveland
Chilham
Harrietsham
Molash
Petham
Broomfield
Kingswood
Lenham
Challock Lees
Challock Forest
Godmersham
Lenham Hth
Charing
Sole Street
Grafty Green
Boughton Malherbe
Bilting
Ulcombe
Charing Heath
Boughton Aluph
Crundale
Waltham
Westwell Leacon
Egerton
Westwell
Boughton Lees
Wye
Egerton Forstal
Gt. Stour
Hassell Street
Elmsted Court
Pluckley
Little Chart
Headcorn
Hothfield
Brook
Hastingleigh
Pluckley Thorne
Kennington
Lymbridge Green
Smarden Bell
ASHFORD
Maltman's Hill
Hinxhill
Brabourne
Stowting
Smarden
Willesborough
Great Chart
Haffenden Quarter
Sevington
Brabourne Lees
Standen
Bethersden
E. Stour
Mersham
Smeeth
Sellindge
Three Chimneys
Kingsnorth
Stanford
Biddenden
Cheeseman's Green
Shadoxhurst
Aldington
High Halden
Redbrook Street
Newingreen
St. Michael's
Orlestone Forest
Bonnington
East End
Lympne
Military Canal
Tenterden
Woodchurch
Ruckinge
Bilsington
Kenardington
Hamstreet
Rolvenden
Shirley Moor
Warehorne
Burmarsh
Royal
Reading Street
Newchurch
Small Hythe
Snave
Romney Marsh
Rolvenden
M2
M20
A2
A20
A28
A249
A251
A252
A262
A274
A290
A299
A2070
B2231
B2163
B2067
B2068
B2069
B2080
B2077

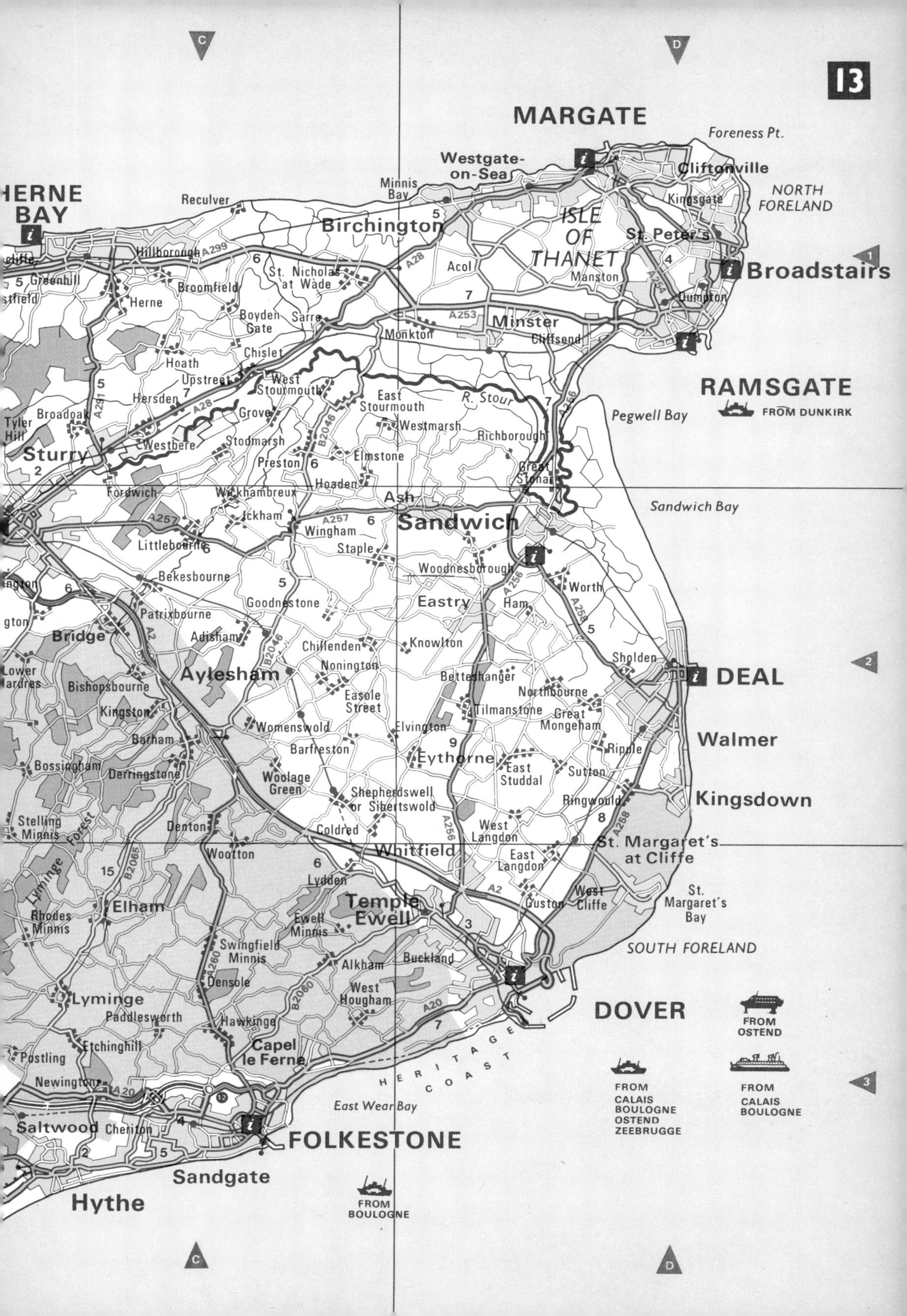

MARGATE

Foreness Pt.

Westgate-on-Sea

Cliftonville

Minnis Bay

Kingsgate

NORTH FORELAND

HERNE BAY

Reculver

Birchington

ISLE OF THANET

St. Peter's

Broadstairs

cliffe

Hillborough A299

St. Nicholas at Wade

Acol

Manston

Dumpton

Greenhill

Broomfield

Herne

A28

stfield

Boyden Gate

Sarre

Monkton

A253

Minster

Cliffsend

RAMSGATE

Tyler Hill

Hoath

Upstreet

West Stourmouth

East Stourmouth

R. Stour

Pegwell Bay

FROM DUNKIRK

Broadoak

Hersden

Grove

Westmarsh

Richborough

Chislet

A28

Sturry

Westbere

Stodmarsh

Preston

Elmstone

Great Stonar

Sandwich Bay

Fordwich

Wickhambreux

Hoaden

Littlebourne

Ickham

Ash

Sandwich

A257

Wingham

Staple

Woodnesborough

Bekesbourne

Goodnestone

Eastry

Ham

Worth

Bridge

Patrixbourne

Knowlton

Sholden

DEAL

Adisham

Chillenden

Betteshanger

Northbourne

Walmer

Aylesham

Nonington

Easole Street

Tilmanstone

Great Mongeham

Kingsdown

Lower Hardres

Bishopsbourne

Womenswold

Barfreston

Elvington

Eythorne

Ripple

Sutton

Ringwould

Kingston

Barham

Woolage Green

East Studdal

St. Margaret's at Cliffe

Bossingham

Derringstone

Shepherdswell or Sibertswold

West Langdon

St. Margaret's Bay

Stelling Minnis

Denton

Coldred

Whitfield

East Langdon

Guston

West Cliffe

SOUTH FORELAND

Elham

Wootton

Lydden

Temple Ewell

A2

Rhodes Minnis

Ewell Minnis

Buckland

DOVER

FROM OSTEND

Lyminge

Swingfield Minnis

Alkham

West Hougham

Paddlesworth

Densole

FROM CALAIS BOULOGNE OSTEND ZEEBRUGGE

FROM CALAIS BOULOGNE

Postling

Etchinghill

Hawkinge

Capel le Ferne

Newington

A20

HERITAGE COAST

East Wear Bay

Saltwood

Cheriton

FOLKESTONE

Sandgate

FROM BOULOGNE

Hythe

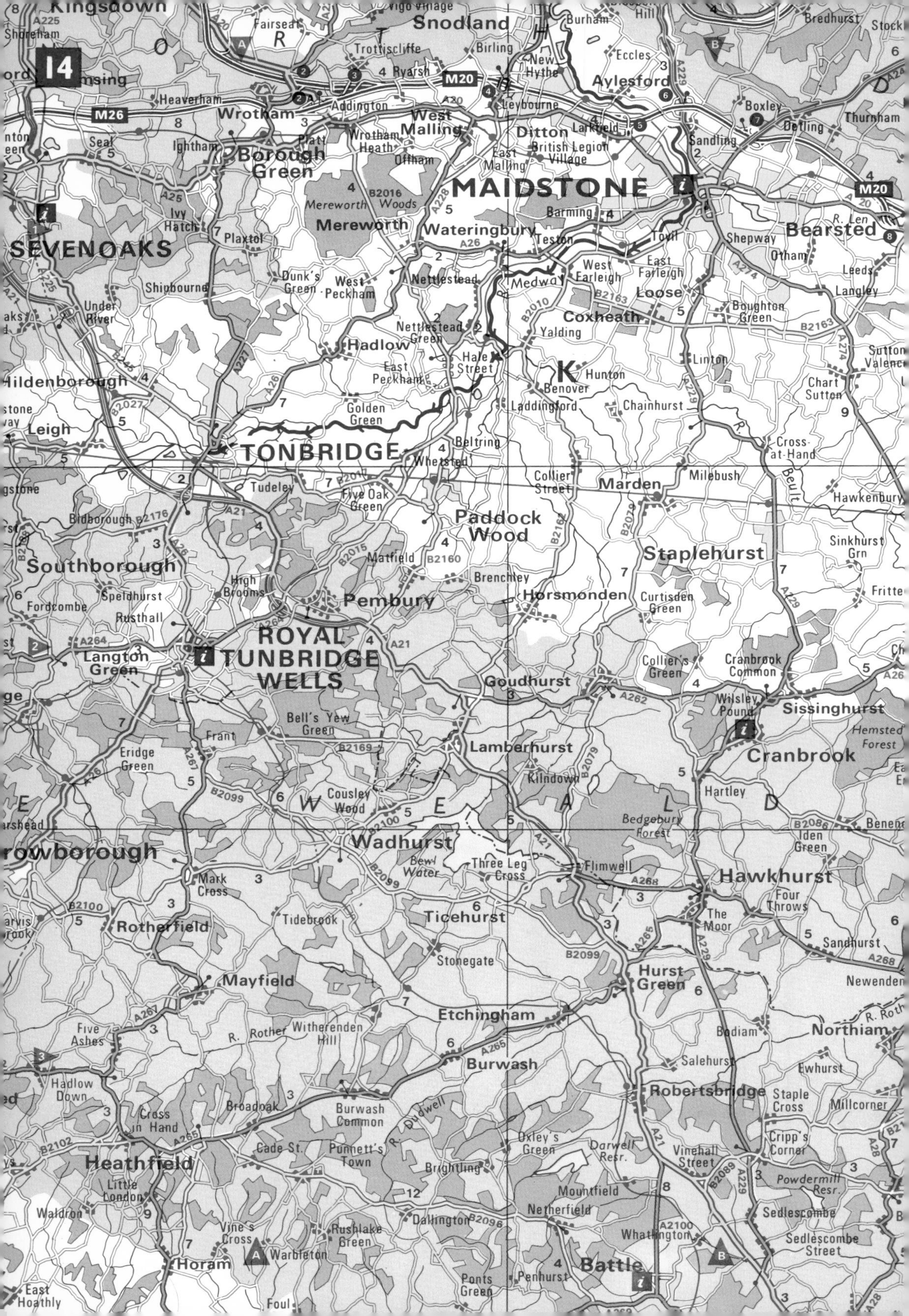

Kingsdown
Shoreham
Fairseat
Vigo Village
Snodland
Burham
Trottiscliffe
Birling
Eccles
Ryarsh
New Hythe
Aylesford
Bredhurst
M20
M26
Heaverham
Wrotham
Addington
West Malling
Leybourne
Larkfield
Boxley
Detling
Thurnham
Seal
Ightham
Platt
Wrotham Heath
Offham
Ditton
British Legion Village
East Malling
Sandling
Borough Green
MAIDSTONE
Mereworth Woods
Mereworth
Wateringbury
Barming
Teston
Tovil
Bearsted
R. Len
Ivy Hatch
Plaxtol
SEVENOAKS
Shipbourne
Dunk's Green
West Peckham
Nettlestead
Medway
West Farleigh
East Farleigh
Shepway
Otham
Leeds
Langley
Loose
Under River
Coxheath
Boughton Green
Hadlow
Nettlestead Green
Yalding
Linton
Sutton Valence
Hale Street
East Peckham
Hunton
Benover
Chart Sutton
Hildenborough
Laddingford
Chainhurst
Golden Green
Leigh
TONBRIDGE
Beltring
Whetsted
Cross-at-Hand
Collier Street
Marden
Milebush
Tudeley
Five Oak Green
Hawkenbury
Bidborough
Paddock Wood
Sinkhurst Grn
Matfield
Staplehurst
Southborough
High Brooms
Brenchley
Speldhurst
Pembury
Horsmonden
Curtisden Green
Fordcombe
Rusthall
ROYAL TUNBRIDGE WELLS
Langton Green
Collier's Green
Cranbrook Common
Goudhurst
Wilsley Pound
Sissinghurst
Bell's Yew Green
Hemsted Forest
Frant
Lamberhurst
Cranbrook
Eridge Green
Kilndown
Hartley
Cousley Wood
Bedgebury Forest
Iden Green
Wadhurst
Bewl Water
Three Leg Cross
Flimwell
Hawkhurst
Crowborough
Mark Cross
Four Throws
Tidebrook
Ticehurst
The Moor
Rotherfield
Sandhurst
Stonegate
Hurst Green
Mayfield
Newenden
Etchingham
R. Rother
Five Ashes
Witherenden Hill
Bodiam
Northiam
Burwash
Ewhurst
Salehurst
Hadlow Down
Robertsbridge
Staple Cross
Millcorner
Cross in Hand
Broadoak
Burwash Common
R. Dudwell
Oxley's Green
Cripp's Corner
Darwell Resr.
Heathfield
Cade St.
Punnett's Town
Vinehall Street
Brightling
Powdermill Resr.
Little London
Mountfield
Sedlescombe
Waldron
Netherfield
Dallington
Vine's Cross
Rushlake Green
Whatlington
Sedlescombe Street
Warbleton
Horam
Battle
Ponts Green
Penhurst
East Hoathly
Foul

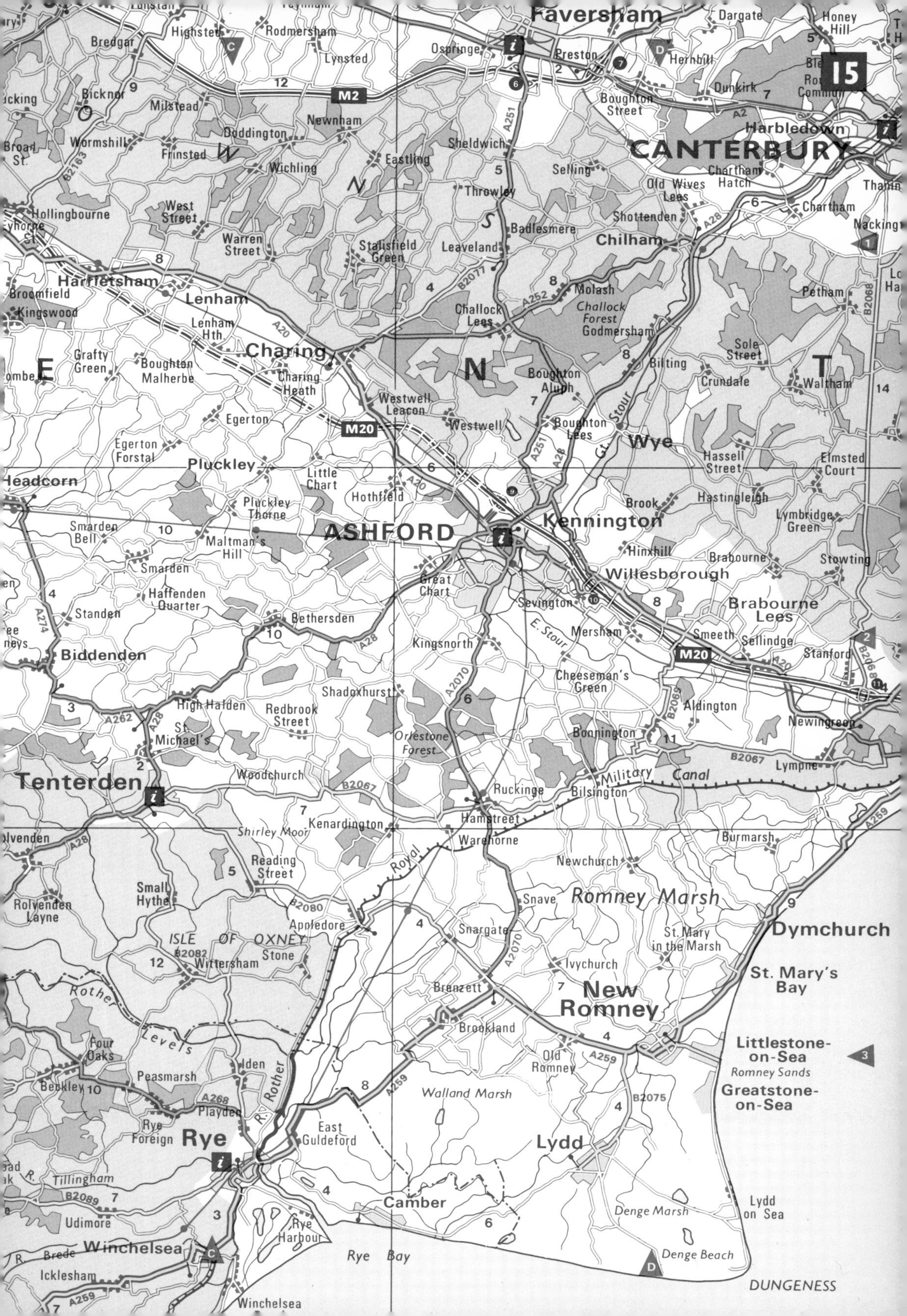
15
Faversham
Dargate
Honey Hill
Highsted
Rodmersham
Bredgar
Lynsted
Ospringe
Preston
Hernhill
Dunkirk
Bicknor
Milstead
M2
Newnham
Boughton Street
Harbledown
A2
CANTERBURY
Broad St.
Wormshill
Doddington
Frinsted
Sheldwich
Eastling
Wichling
Selling
Chartham Hatch
Old Wives Lees
Throwley
Thanington
Hollingbourne
West Street
Shottenden
Chartham
Badlesmere
Nackington
Warren Street
Stalisfield Green
Leaveland
Chilham
A28
A251
B2163
B2077
Harrietsham
Lenham
Molash
Petham
Broomfield
Kingswood
Challock Lees
Challock Forest
Godmersham
A252
B2068
Lenham Hth.
Charing
A20
Grafty Green
Boughton Malherbe
Charing Heath
Sole Street
Bilting
Crundale
Waltham
Boughton Aluph
E
N
T
Westwell Leacon
Westwell
Egerton
Boughton Lees
Gt. Stour
Wye
M20
Egerton Forstal
Pluckley
Little Chart
Hassell Street
Elmsted Court
Headcorn
Hothfield
Hastingleigh
Pluckley Thorne
Brook
Lymbridge Green
Smarden Bell
ASHFORD
Kennington
Maltman's Hill
Hinxhill
Brabourne
Stowting
Smarden
Willesborough
Great Chart
Haffenden Quarter
Sevington
Brabourne Lees
Standen
Bethersden
Mersham
E. Stour
Smeeth
A274
Kingsnorth
Sellindge
Stanford
Biddenden
Cheeseman's Green
A2070
Shadoxhurst
Aldington
High Halden
Redbrook Street
B2069
Newingreen
A262
St. Michael's
Orlestone Forest
Bonnington
B2067
Lympne
Tenterden
Woodchurch
Military Canal
Ruckinge
Bilsington
Hamstreet
Kenardington
Shirley Moor
Warehorne
Rolvenden
Royal
Burmarsh
A259
Reading Street
Newchurch
Small Hythe
Snave
Romney Marsh
Rolvenden Layne
B2080
Appledore
Dymchurch
Snargate
St. Mary in the Marsh
ISLE OF OXNEY
Stone
B2082
Wittersham
Ivychurch
St. Mary's Bay
Rother
Brenzett
New Romney
Levels
Brookland
Four Oaks
Old Romney
Littlestone-on-Sea
Iden
Romney Sands
Peasmarsh
R. Rother
Beckley
A268
Walland Marsh
B2075
Greatstone-on-Sea
Playden
Rye Foreign
Rye
East Guldeford
Lydd
R. Tillingham
B2089
Udimore
Camber
Lydd on Sea
Denge Marsh
Rye Harbour
Winchelsea
R. Brede
Rye Bay
Denge Beach
Icklesham
Winchelsea
DUNGENESS

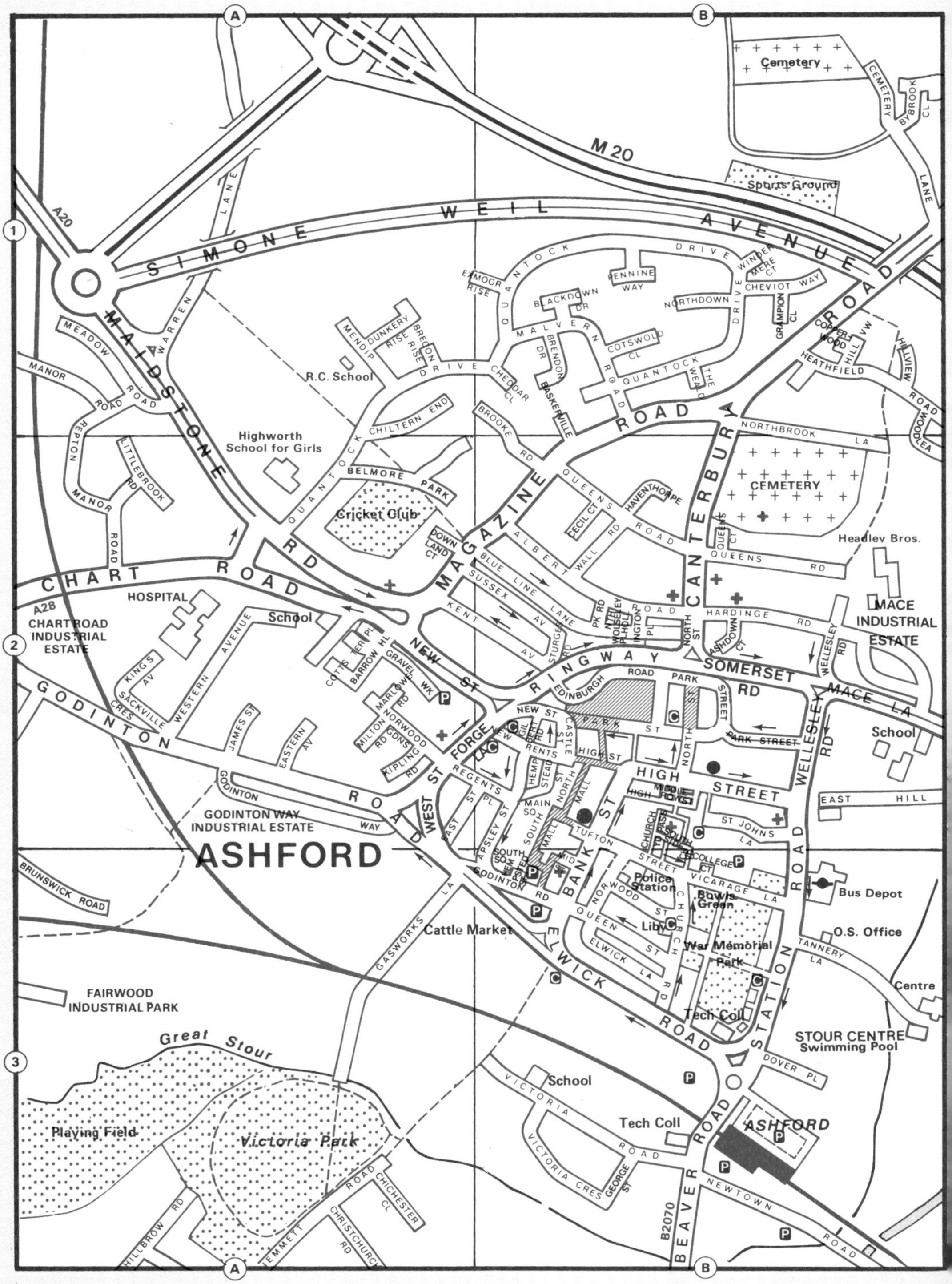

ASHFORD
Road Map page 12   Population 39,994
For extended coverage of this area   —   see Estate Publications RED BOOK—Ashford
Reproduction prohibited without prior permission   © Estate Publications

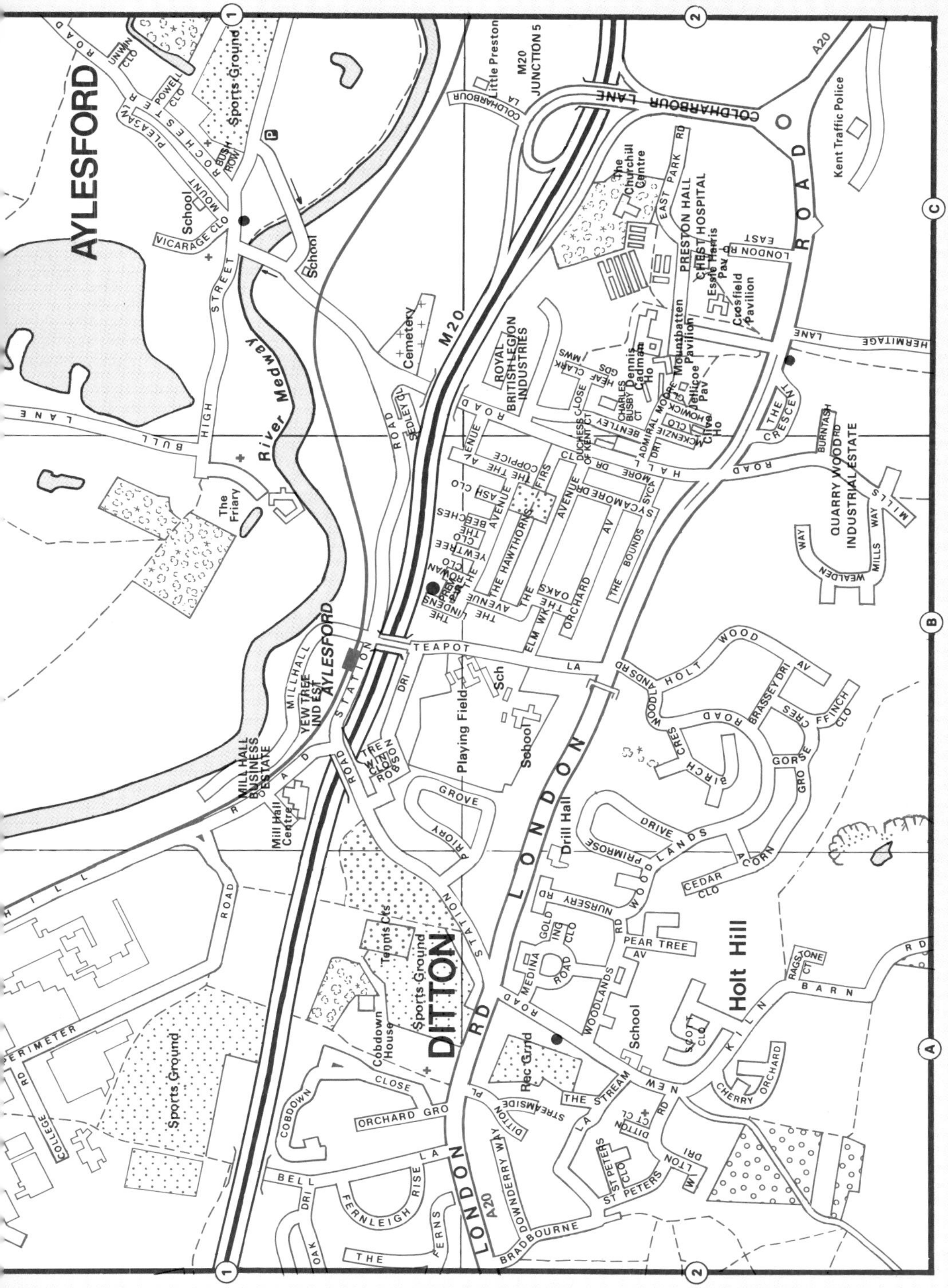

AYLESFORD Population 8,096 DITTON Population 5,168

Road Map page 11

or extended coverage of this area   —   see Estate Publications RED BOOK—Maidstone

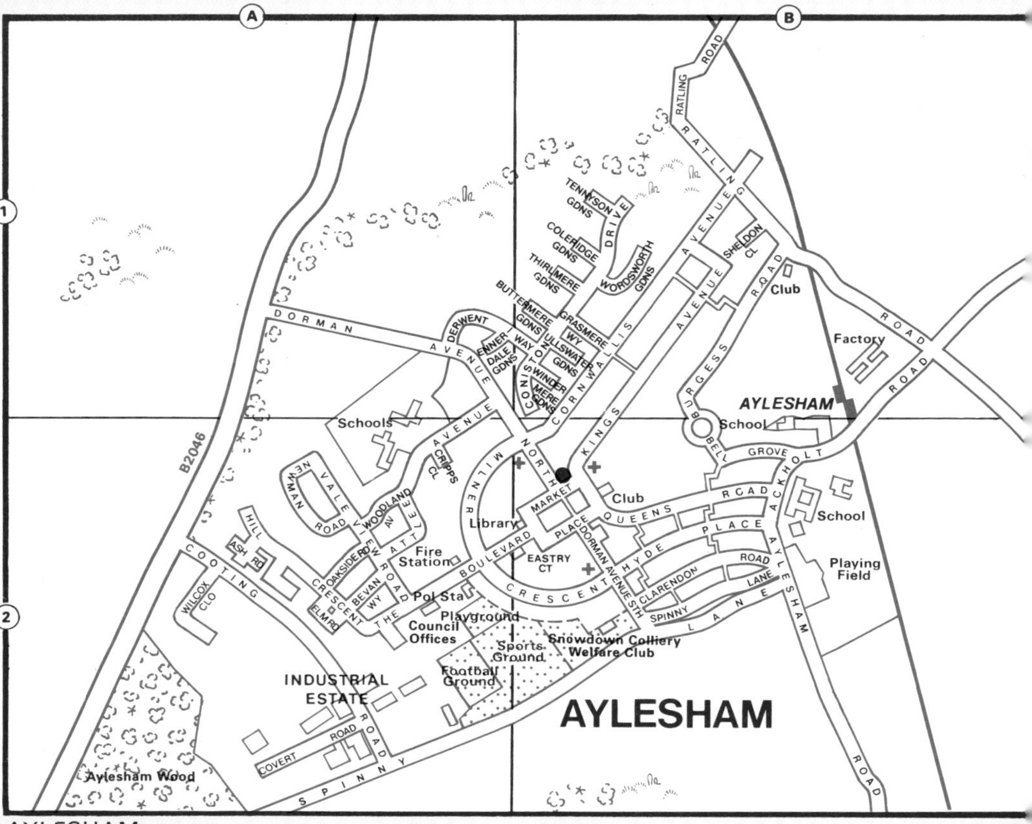

AYLESHAM
Road Map page 13   Population 3,899
For extended coverage of this area—see Estate Publications RED BOOK—Folkestone/Dover

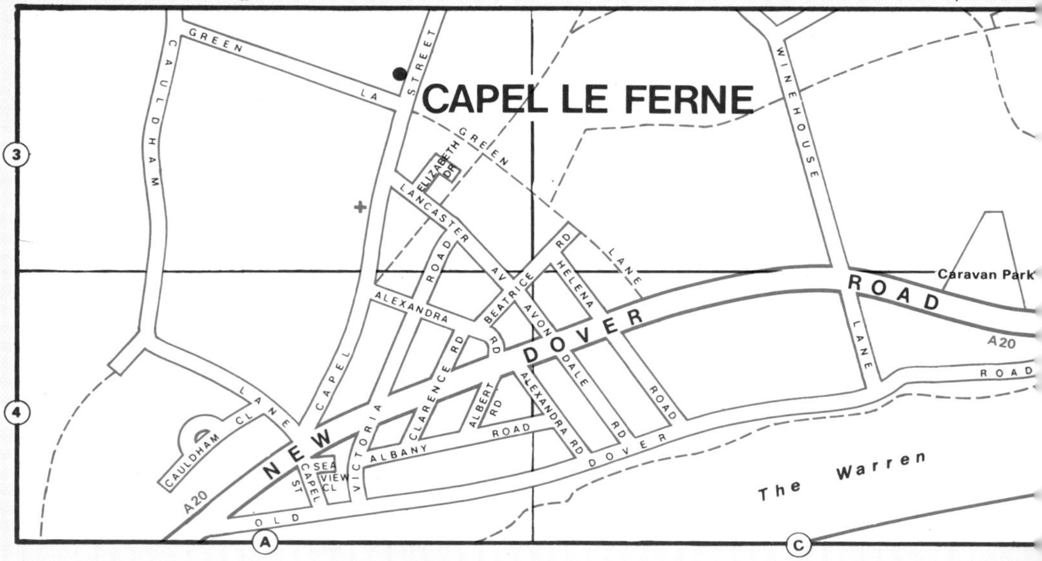

CAPEL-LE-FERNE   # CAPEL-LE-FERNE   Scale 6 inches to 1 mil
Road Map page 13   Population 1,760
For extended coverage of this area—see Estate Publications RED BOOK—Folkestone/Dover

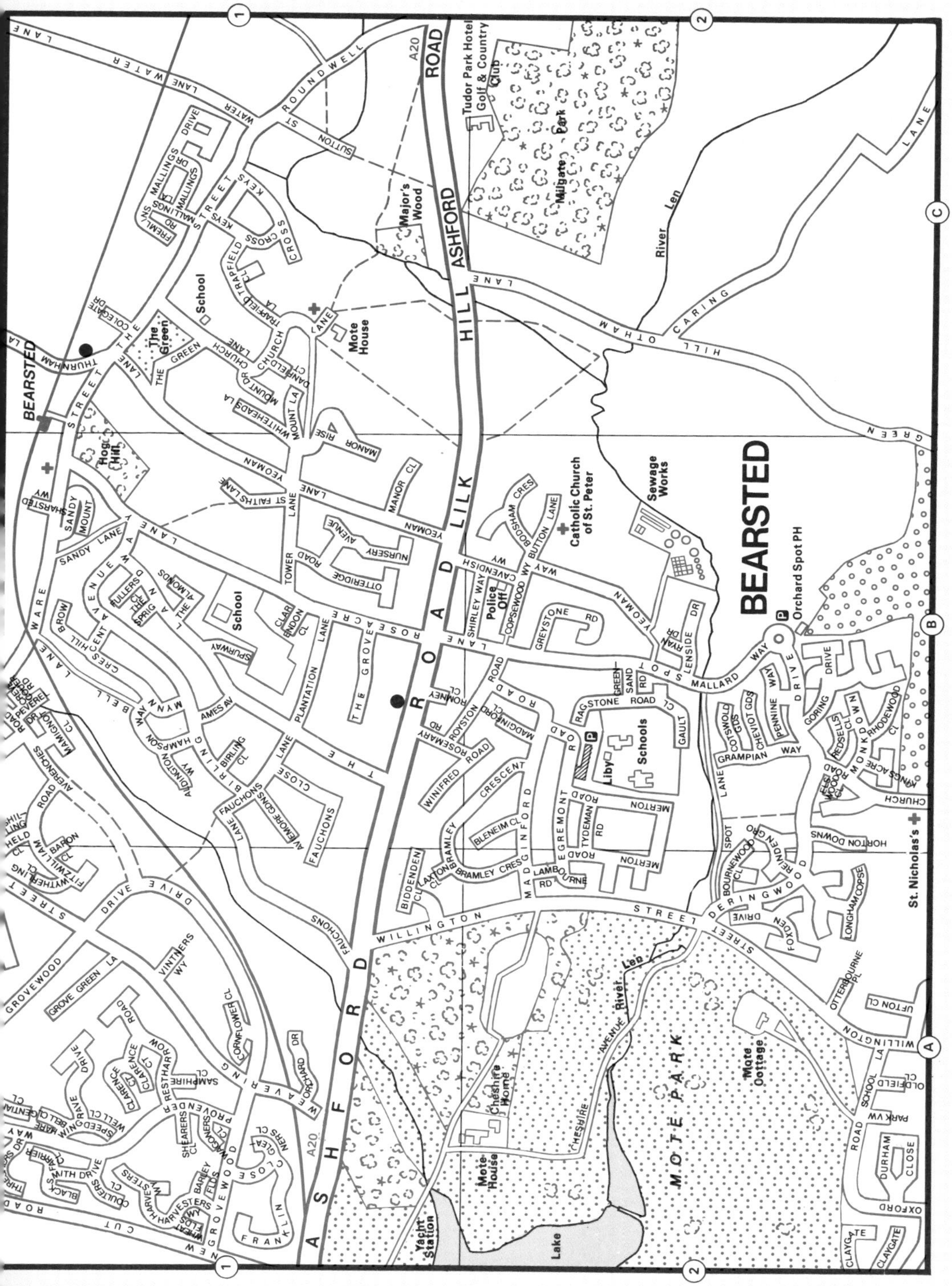

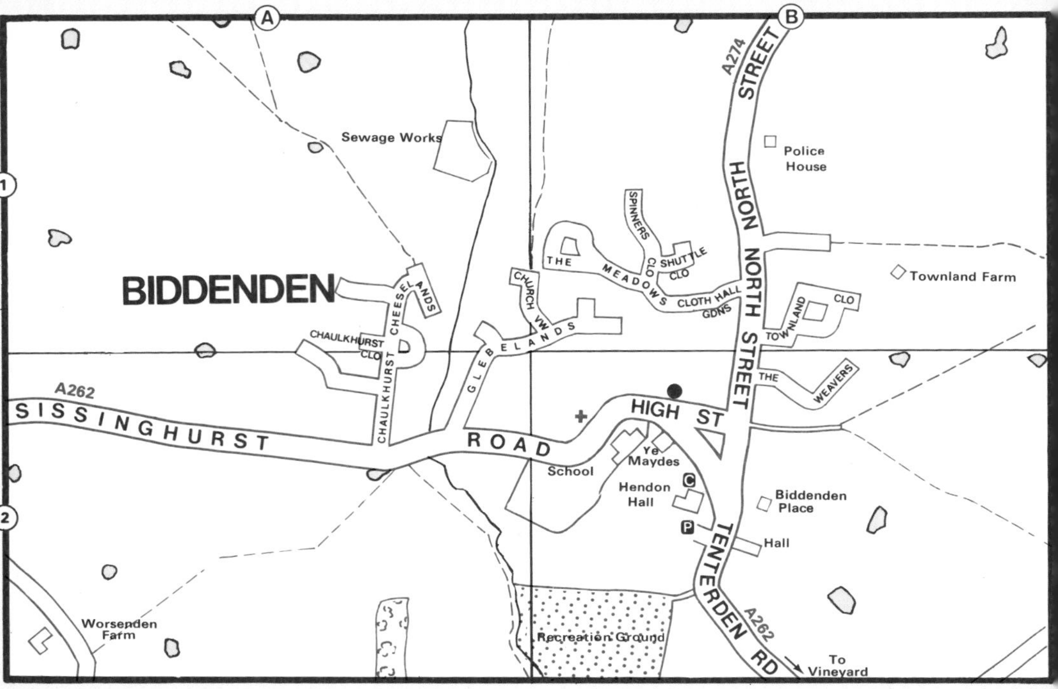

BIDDENDEN
Road Map page 15   Population 2,229

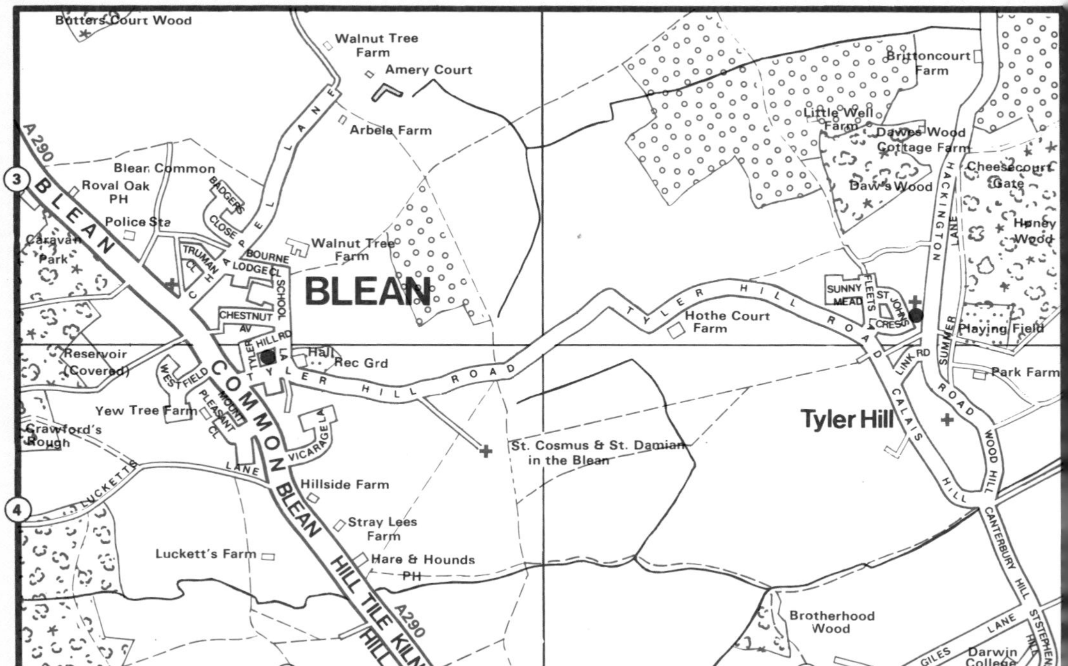

BLEAN
Road Map page 12   Population 1,859

BLEAN

Scale 3 inches to 1 mile

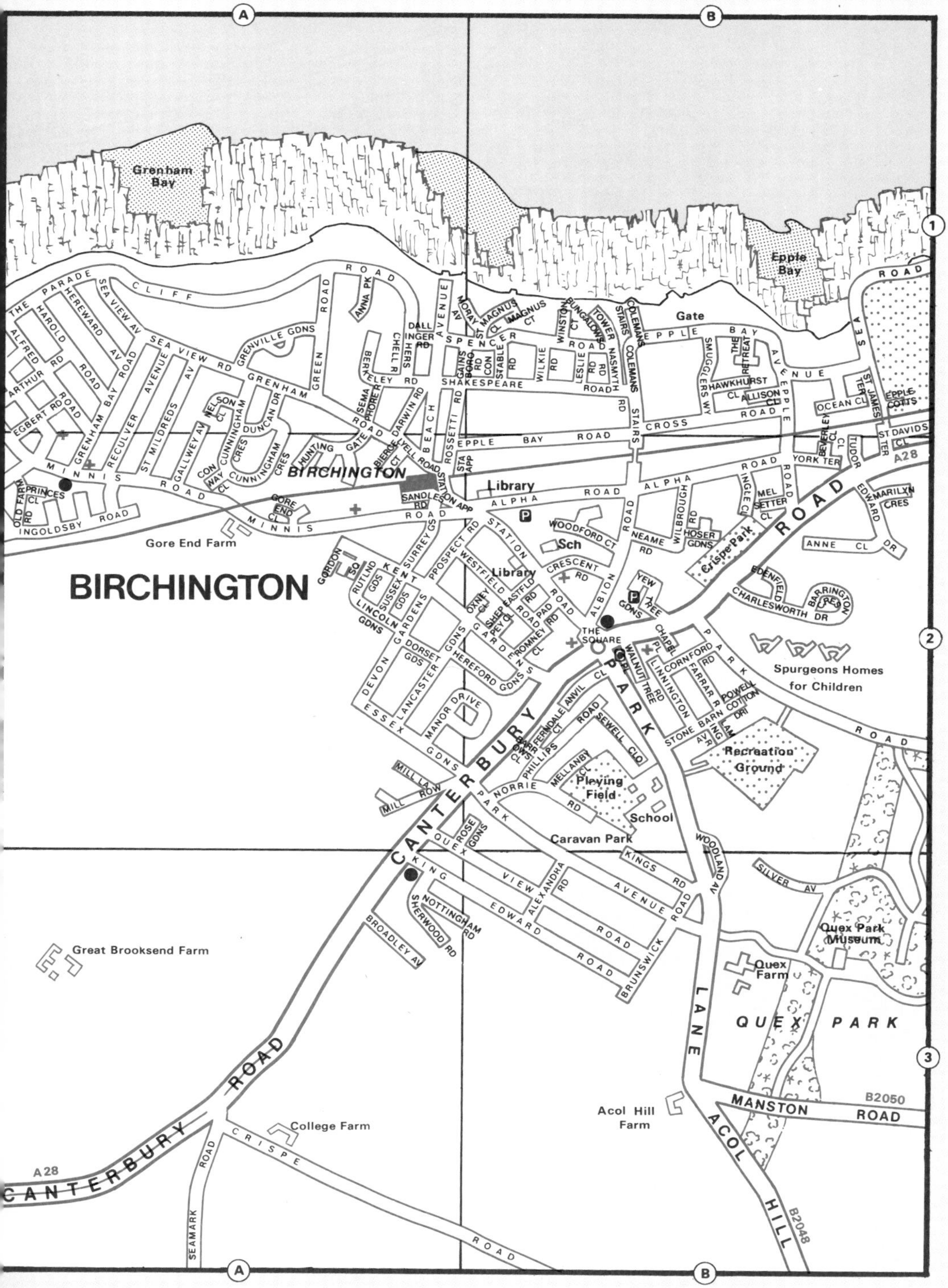

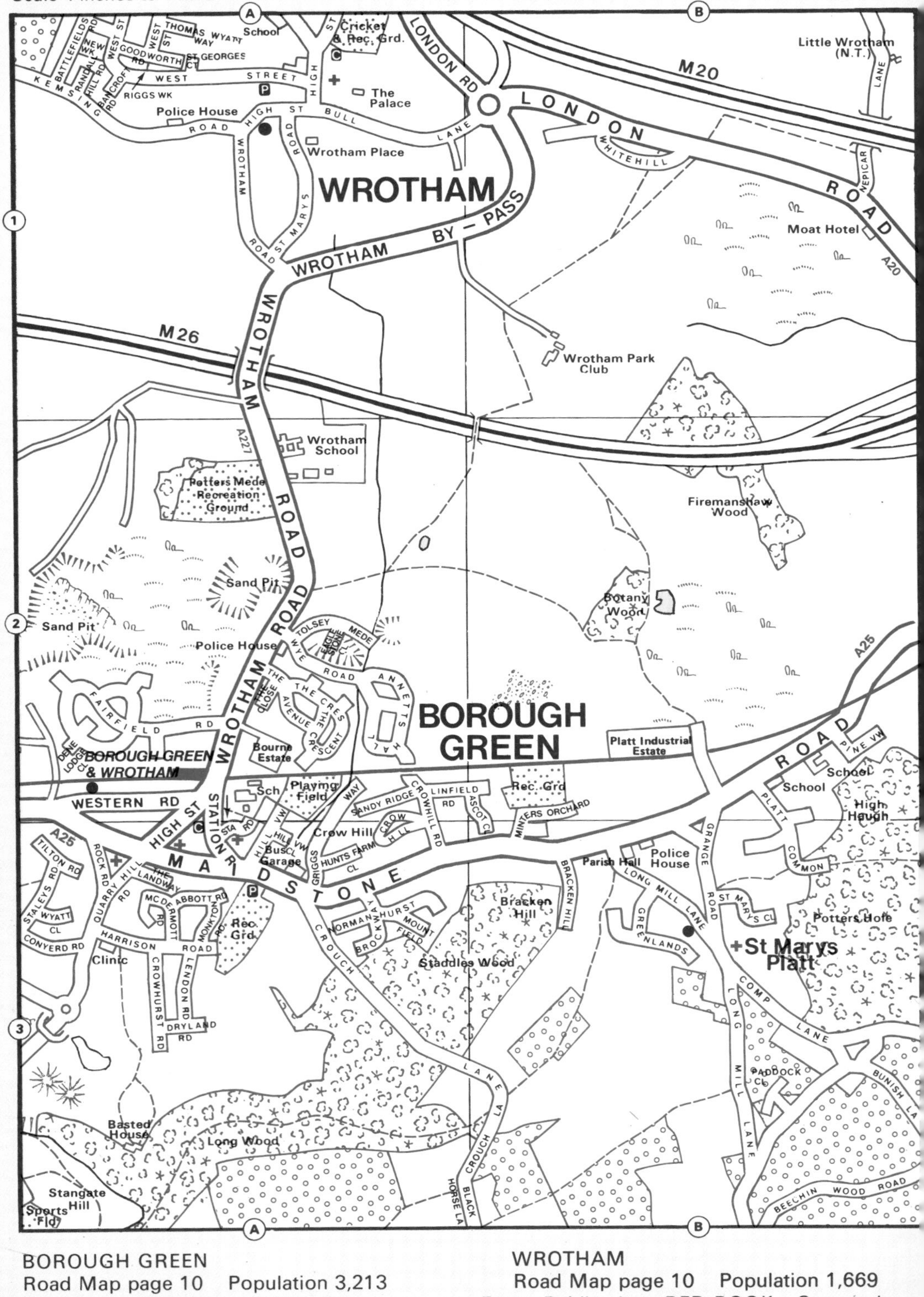

BOROUGH GREEN
Road Map page 10   Population 3,213
For extended coverage of this area   —
Reproduction prohibited without prior permission

WROTHAM
Road Map page 10   Population 1,669
see Estate Publications RED BOOK—Sevenoaks
© Estate Publications

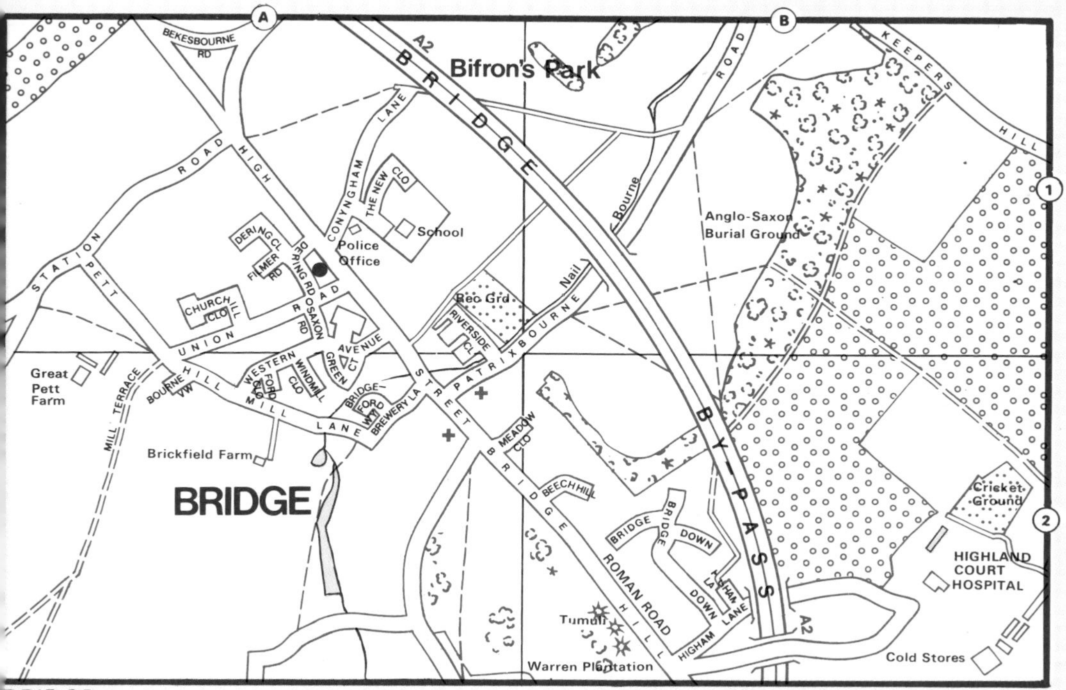

BRIDGE
Road Map page 13   Population 1,302

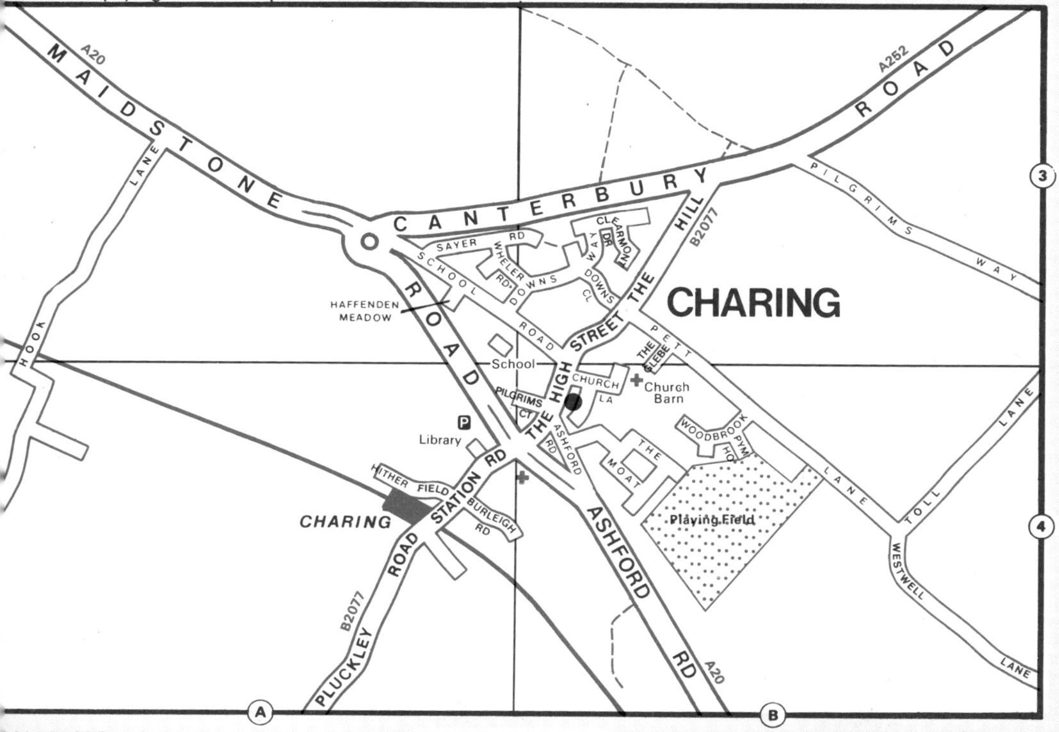

CHARING

Scale 4 inches to 1 mile

For extended coverage of this area   —   see Estate Publications RED BOOK—Ashford

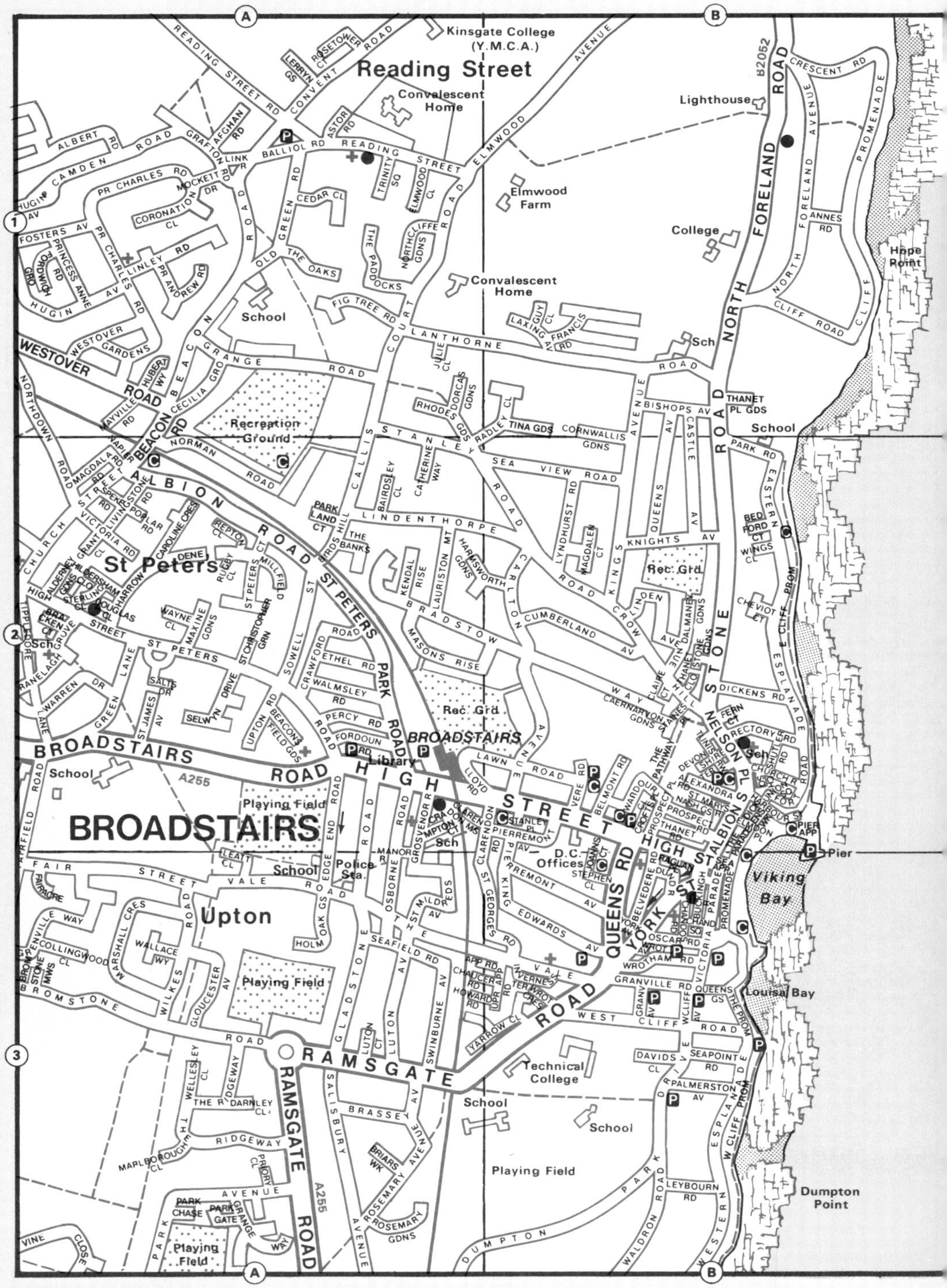

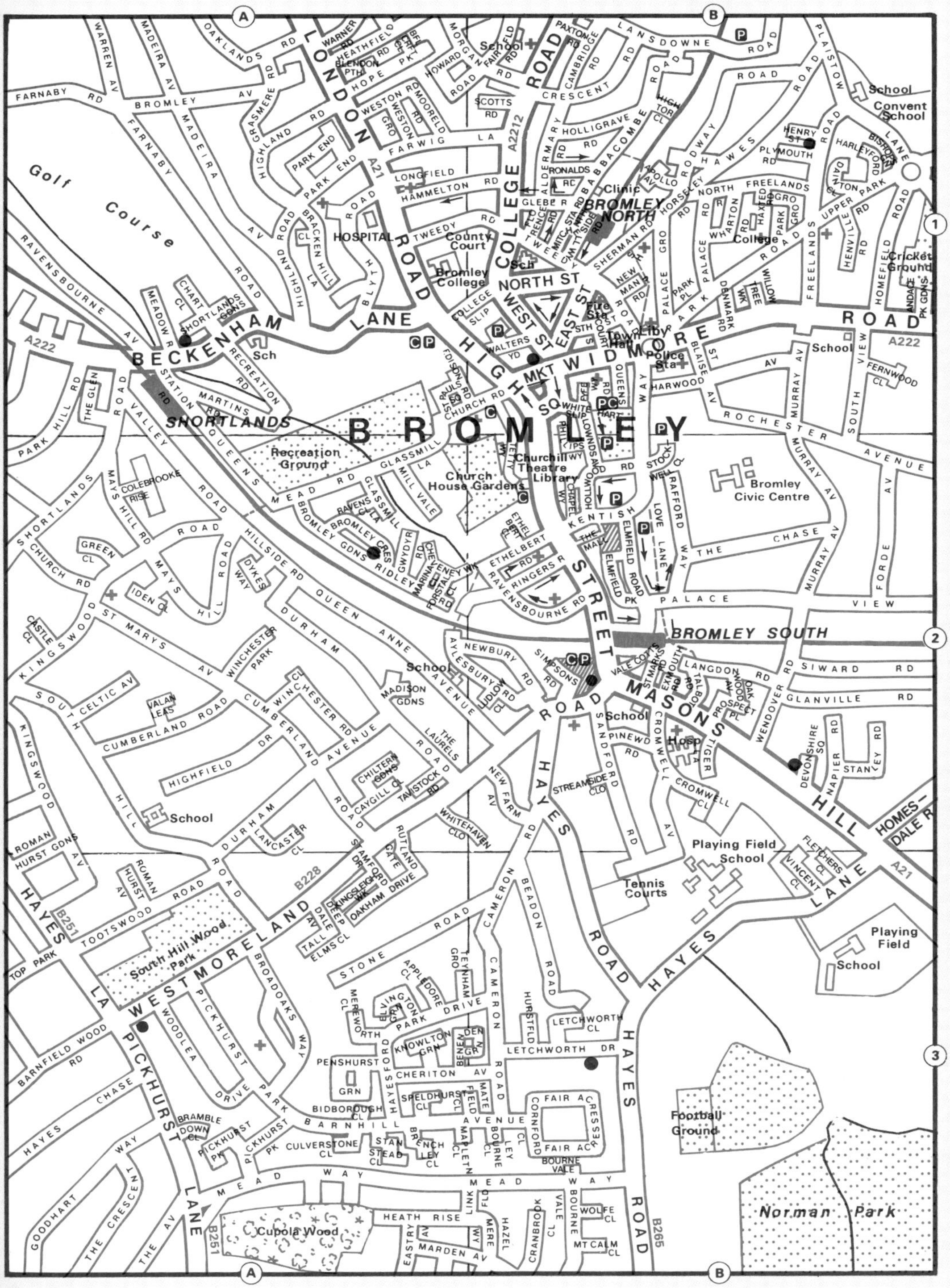

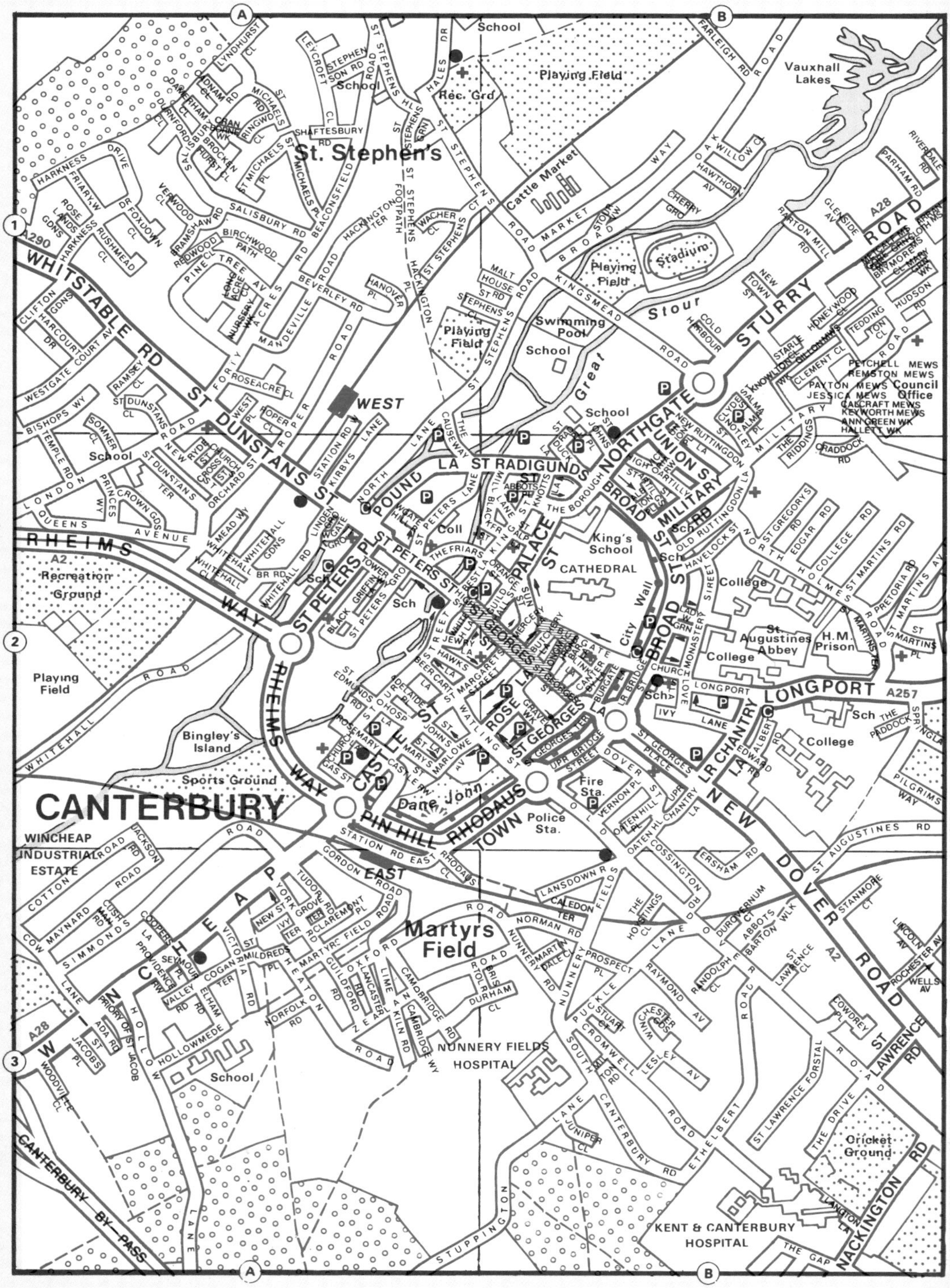

CANTERBURY
Road Map page 12   Population 35,429
For extended coverage of this area—see Estate Publications RED BOOK—Thanet/Canterbury
Reproduction prohibited without prior permission   © Estate Publications

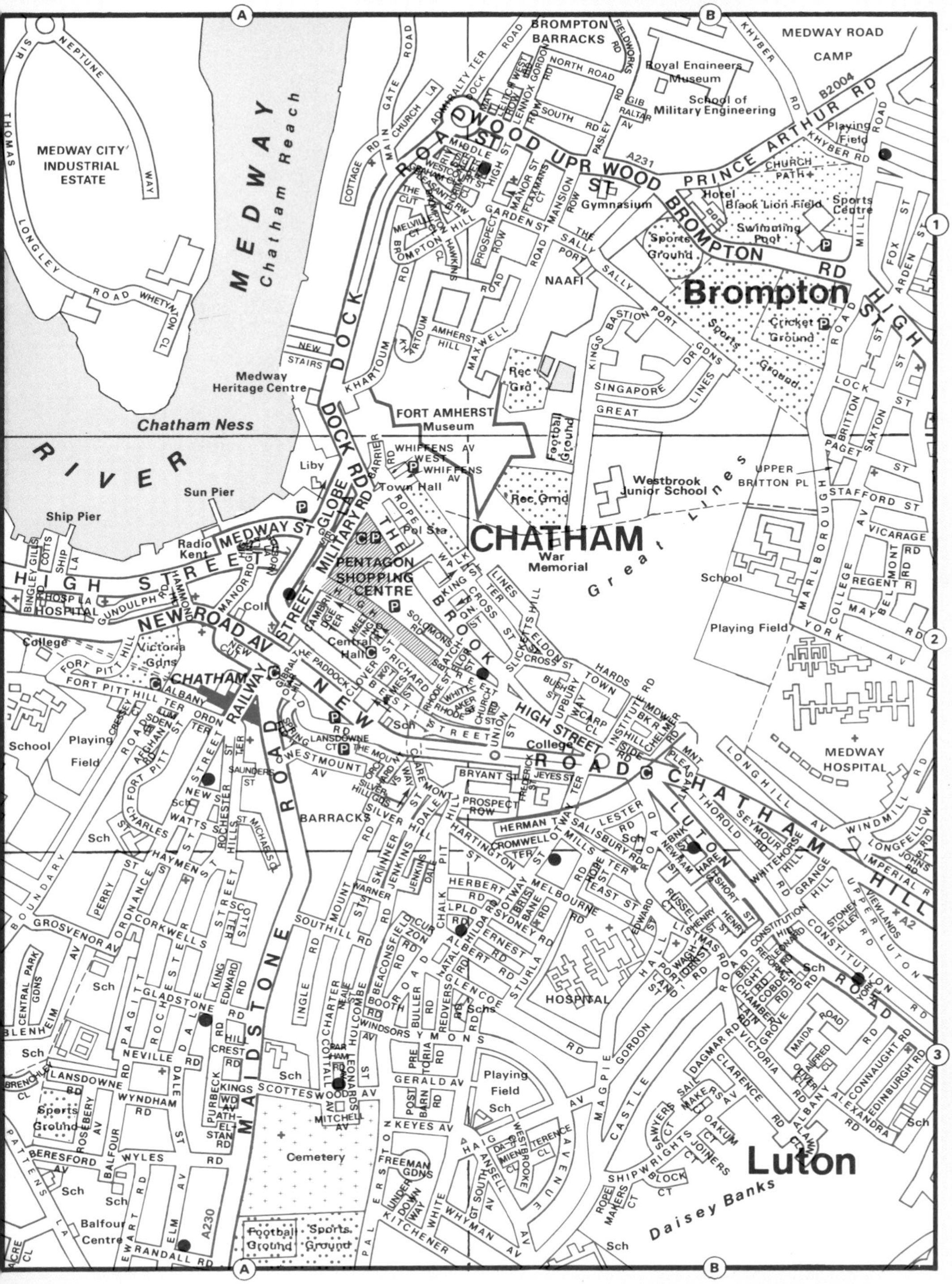

CHATHAM
Road Map page 11   Population 43,915
For extended coverage of this area —see Estate Publications RED BOOK—Medway/Gillingham

CLIFFE Population 4,748
Road Map page 11

For extended coverage of this area—
see Estate Publications RED BOOK —Medway/Gillingham

CLIFFE WOODS
Road Map page 11

CLIFFE WOODS   Scale 4 inches to 1 mile

For extended coverage of this area—see Estate Publications RED BOOK—Medway/Gillingham

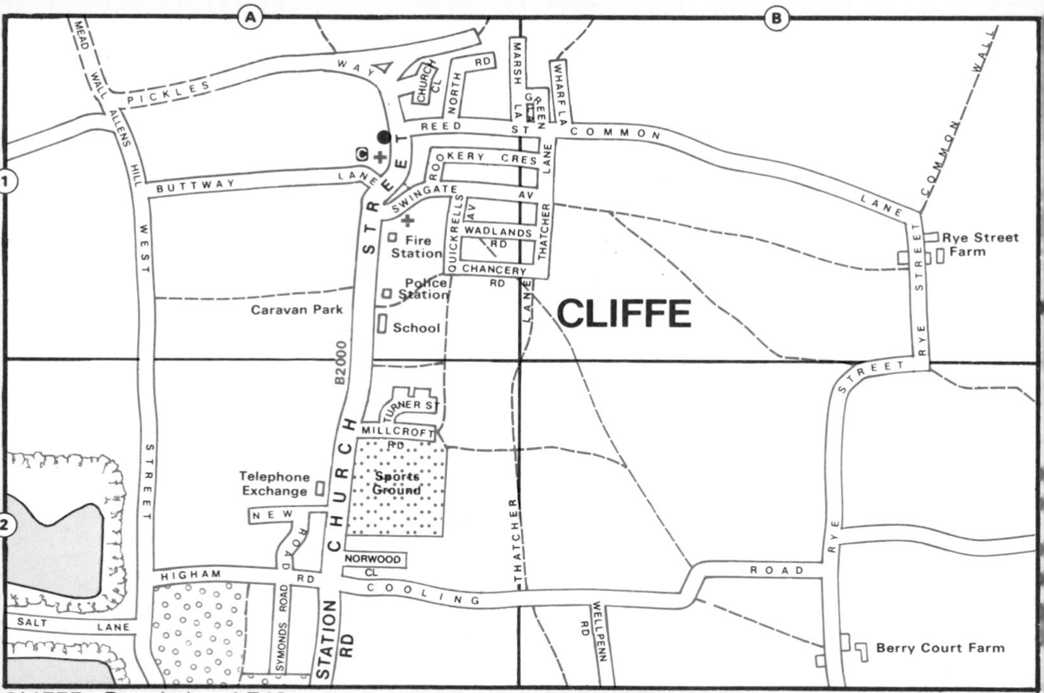

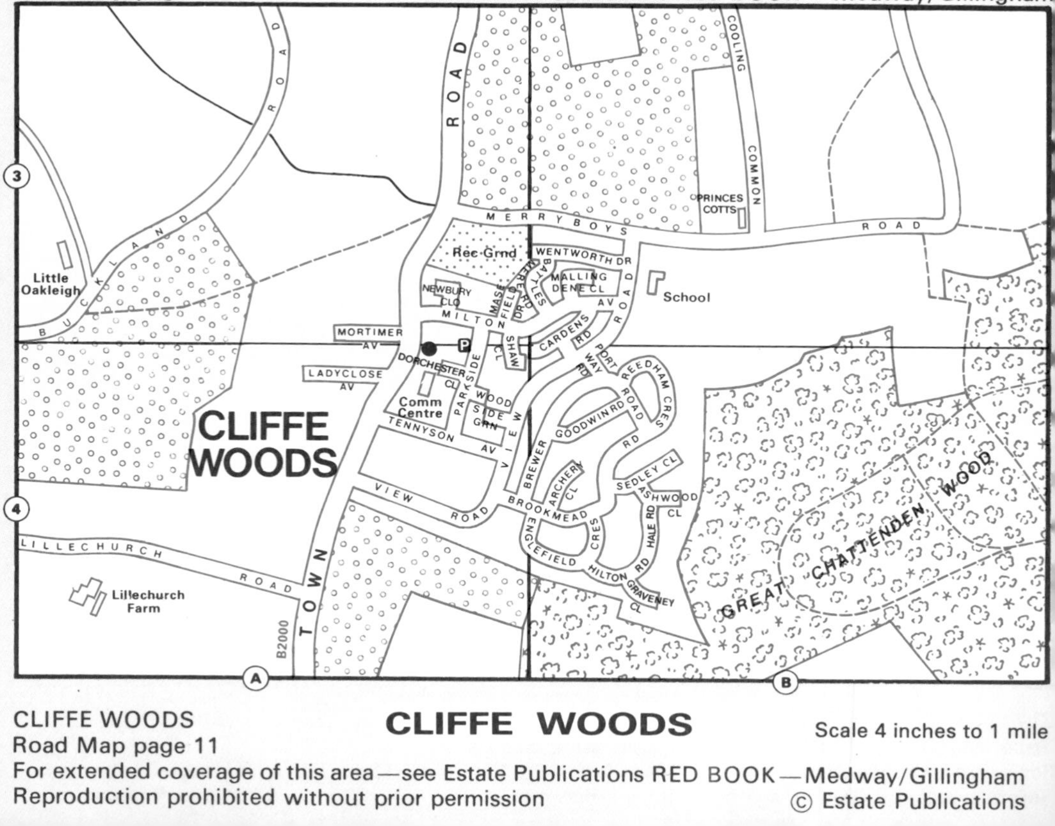

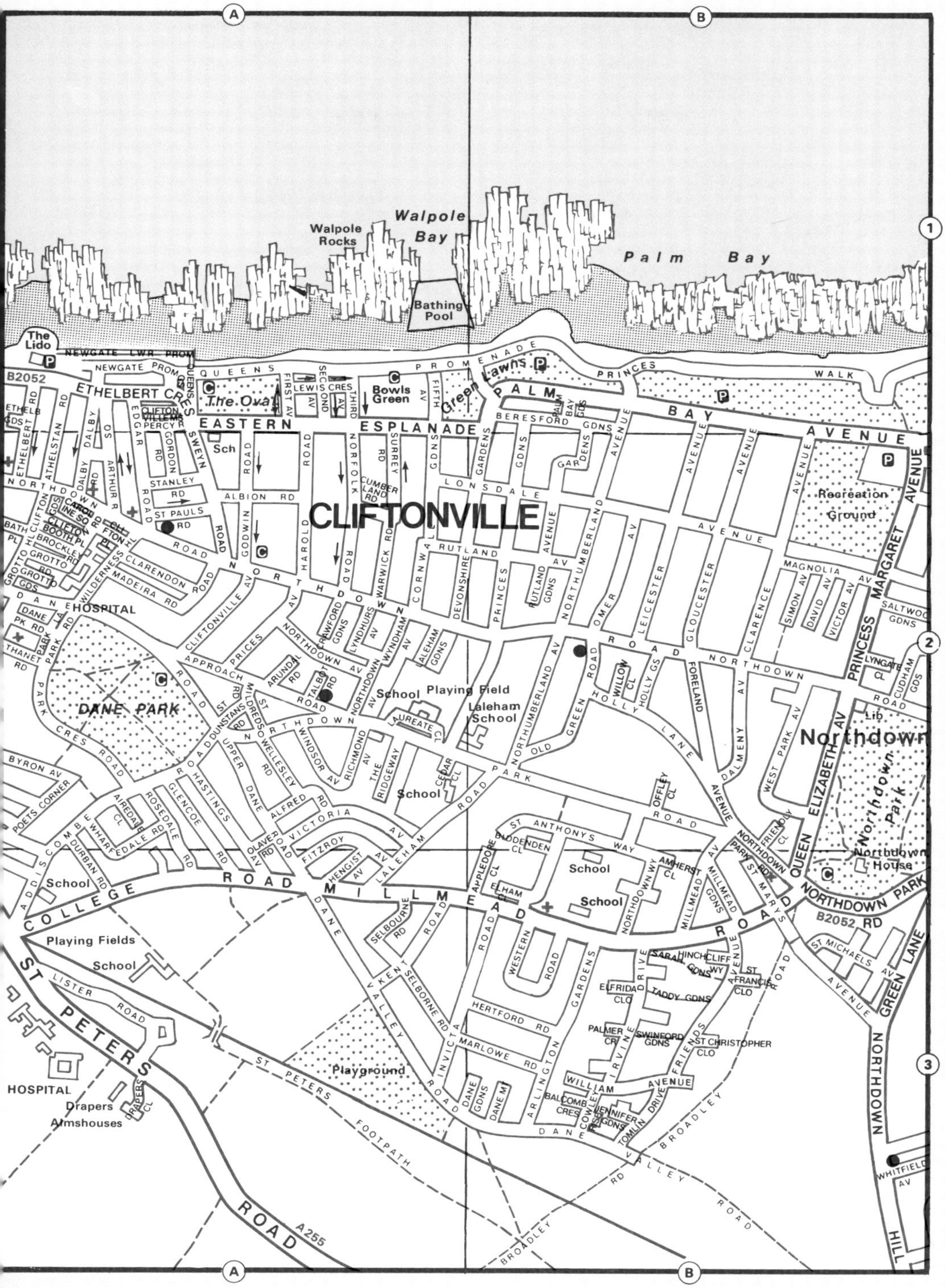

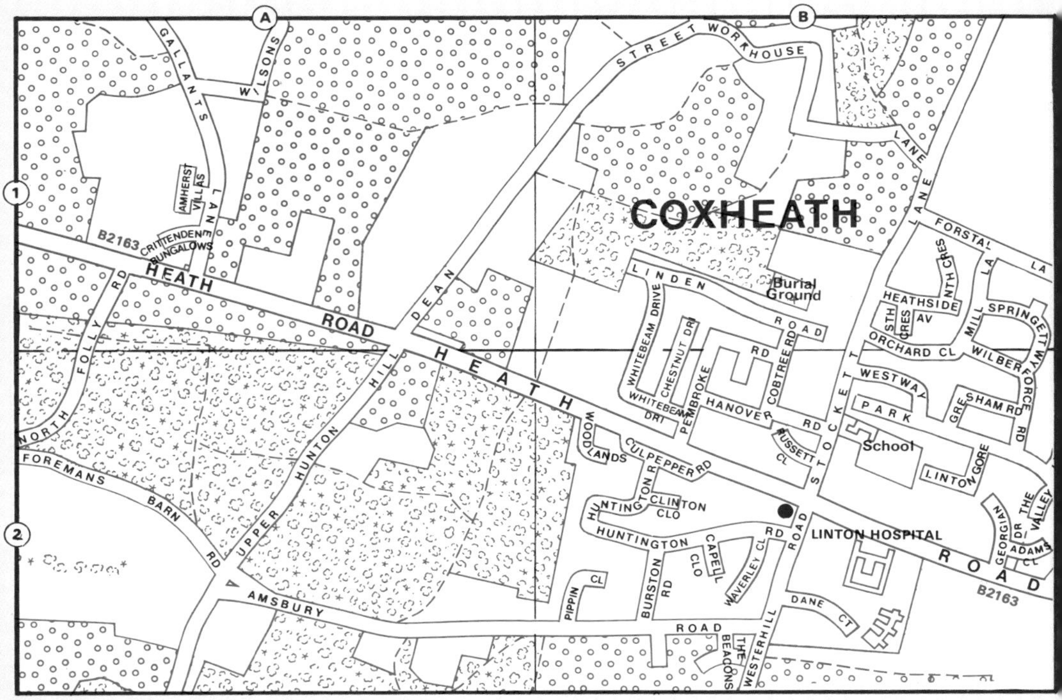

COXHEATH
Road Map page 11   Population 4,230

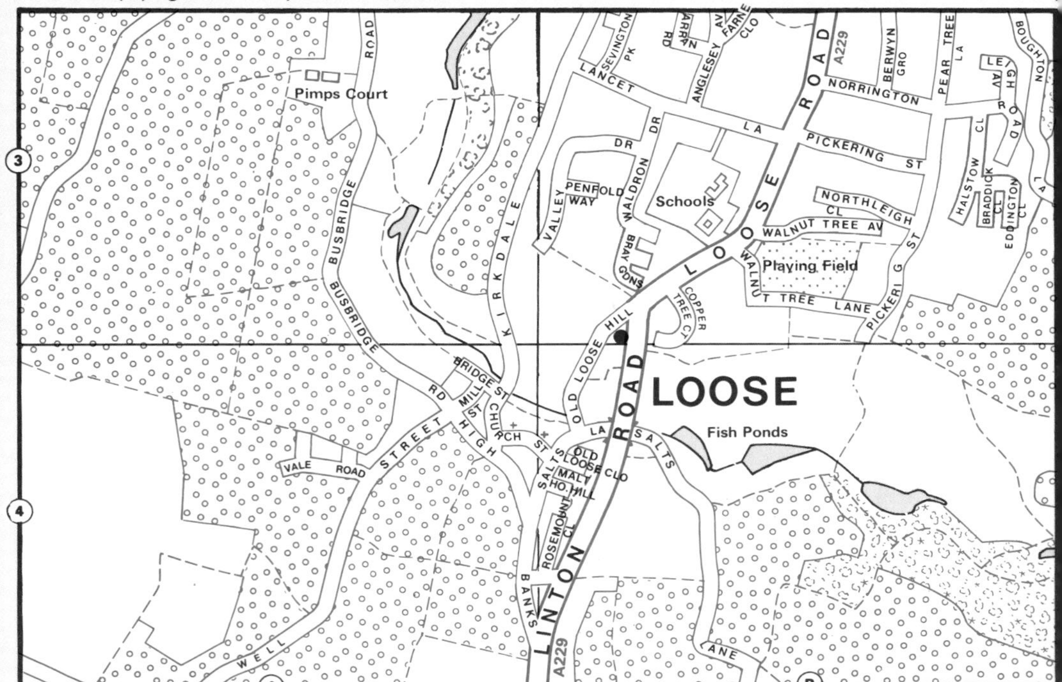

LOOSE
Road Map page 11   Population 2,459

For extended coverage of this area — see Estate Publications RED BOOK—Maidstone

LOOSE   Scale 4 inches to 1 mile

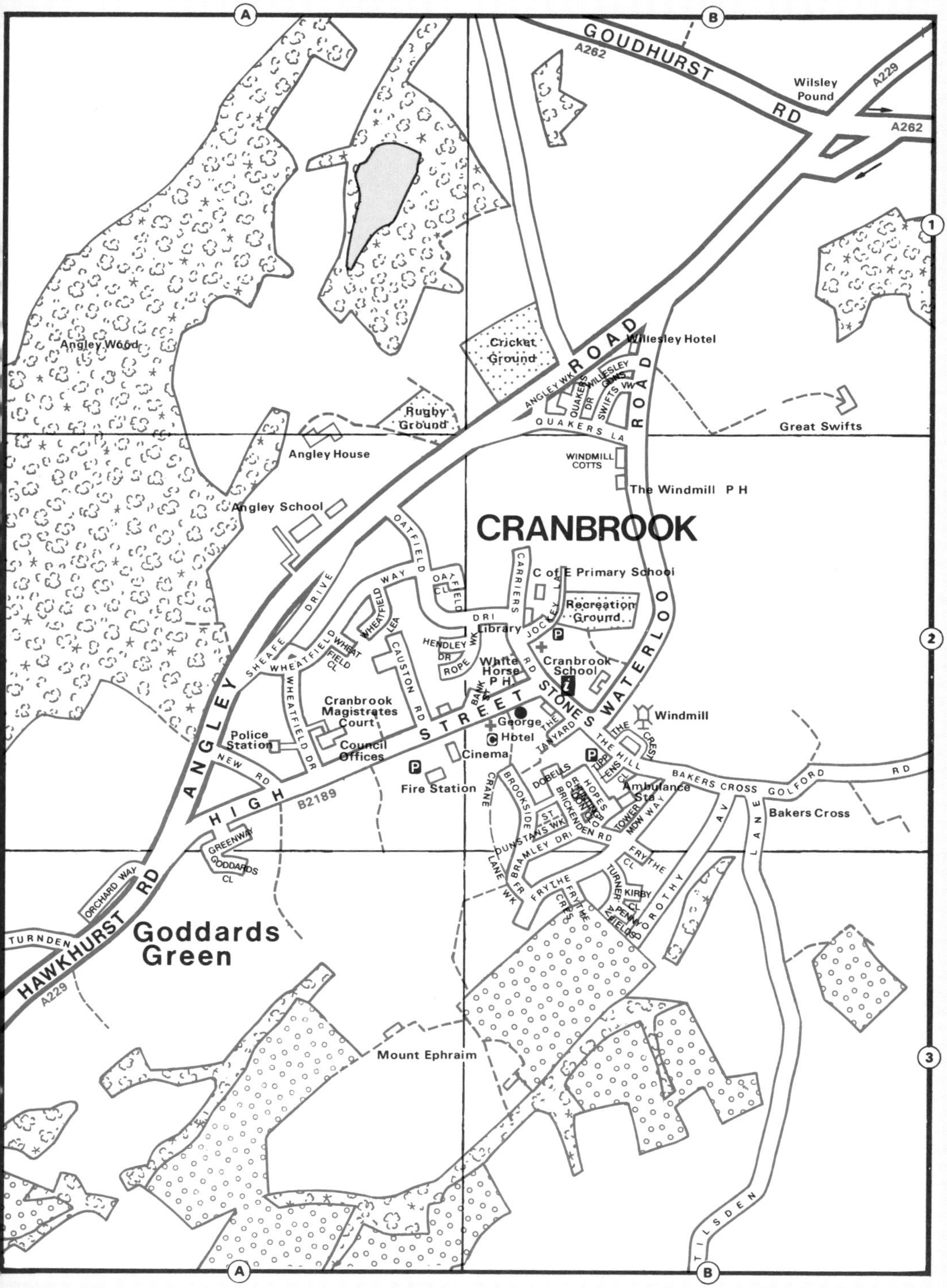

For extended coverage of this area
see Estate Publications RED BOOK—Tunbridge Wells/Tonbridge

For extended coverage of this area— see Estate
Publications RED BOOK—Medway/Gillingham

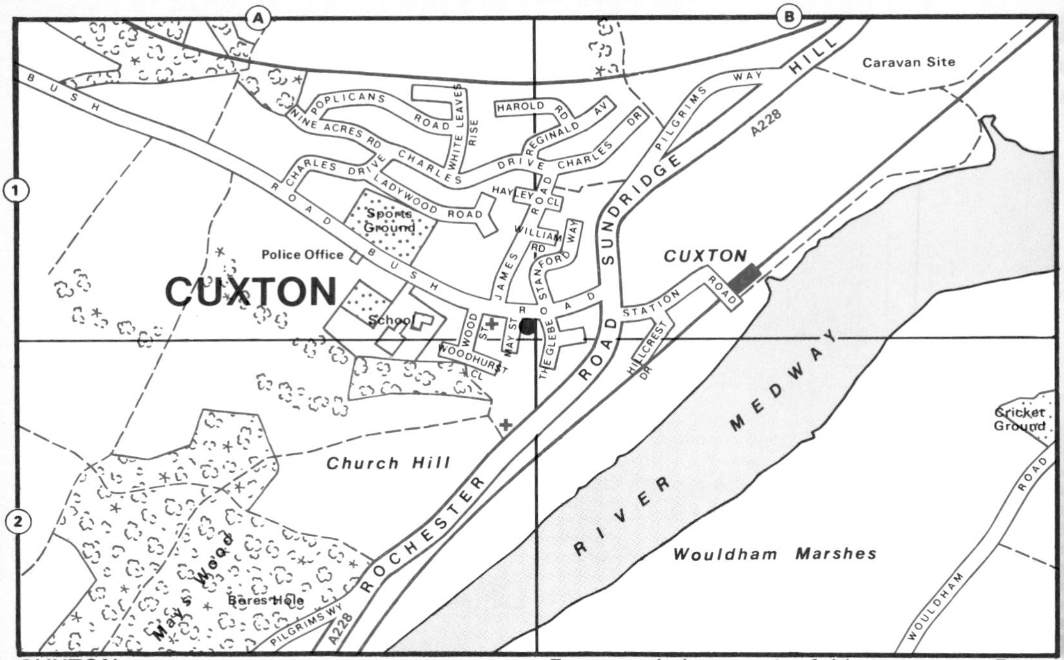

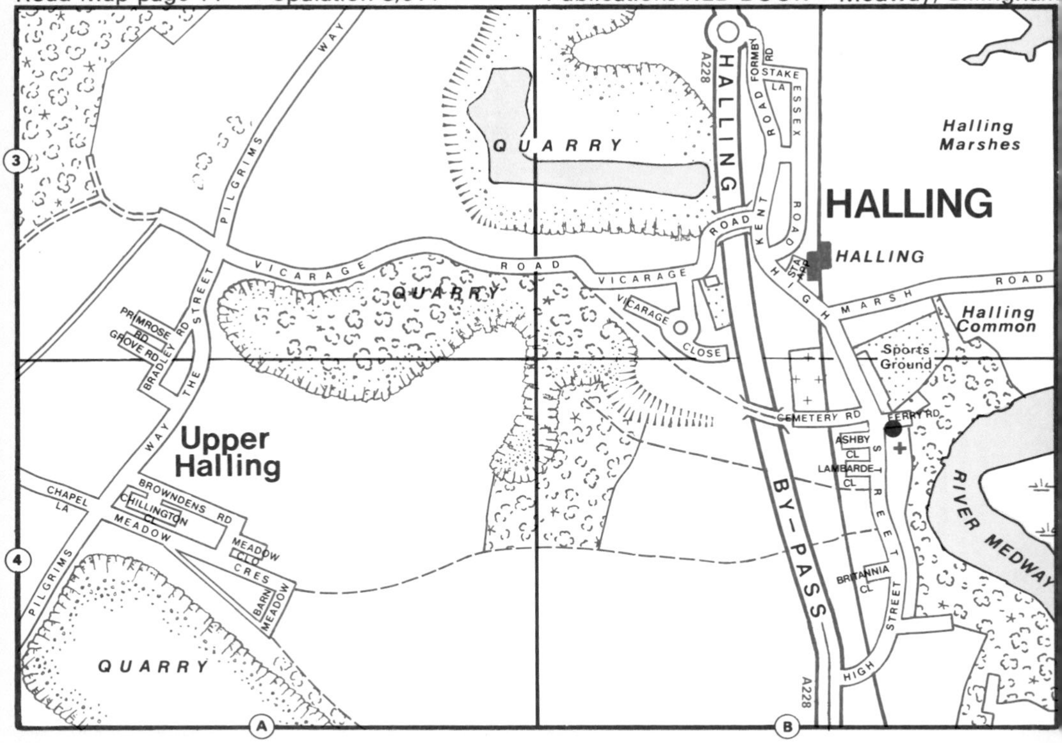

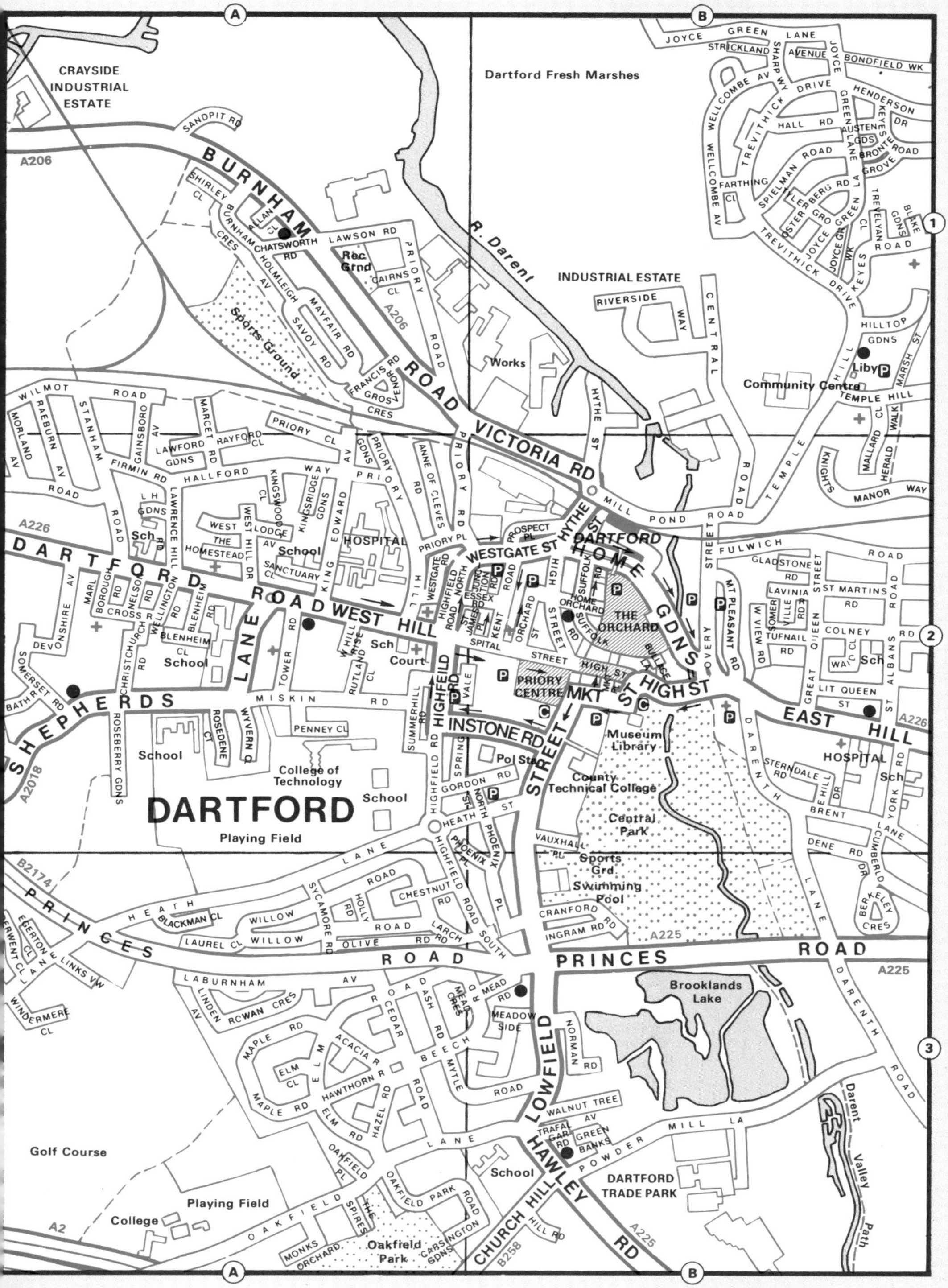

DARTFORD
Road Map page 10   Population 44,003
For extended coverage of this area—see Estate Publications RED BOOK—Gravesend/Dartford
Reproduction prohibited without prior permission   © Estate Publications

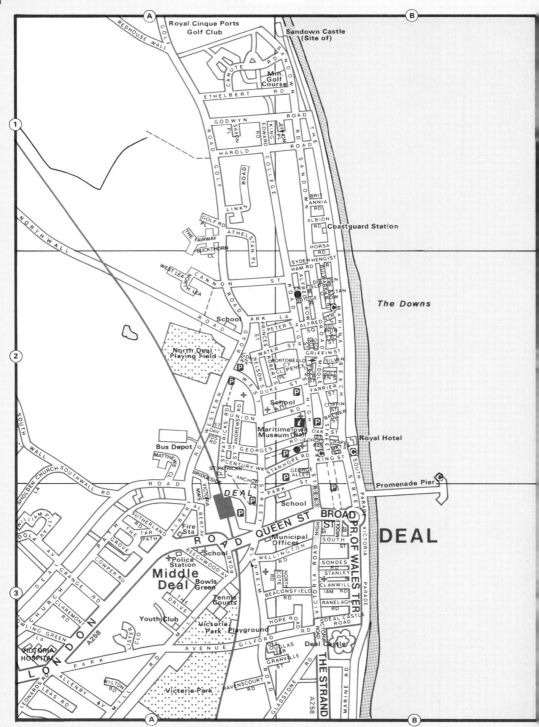

DEAL
Road Map page 13   Population 26,109
For extended coverage of this area—see Estate Publications RED BOOK—Folkestone/Dover

© Estate Publications

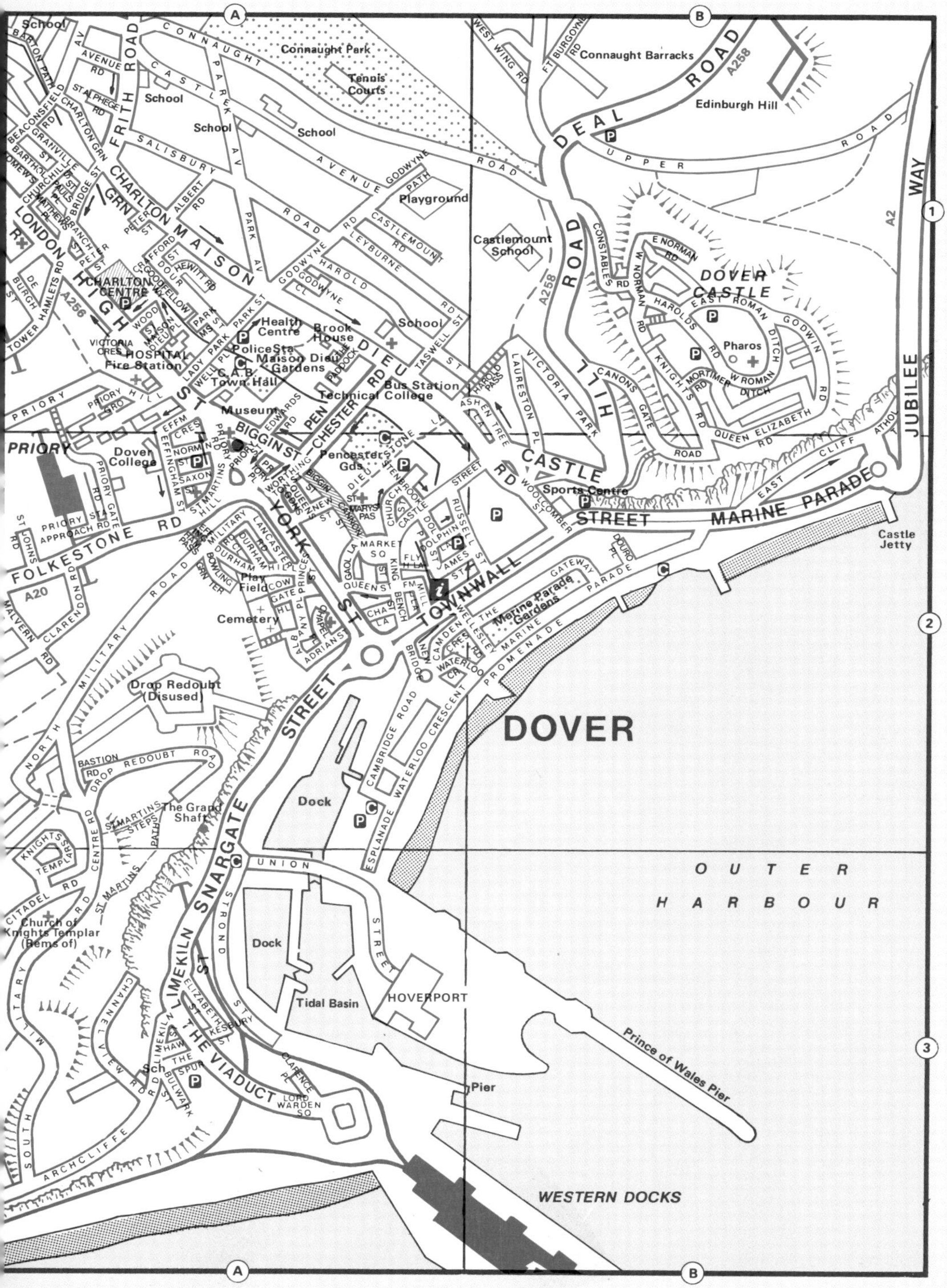

DOVER

OUTER HARBOUR

WESTERN DOCKS

Road Map page 13   Population 32,830

Scale 4 inches to 1 mile

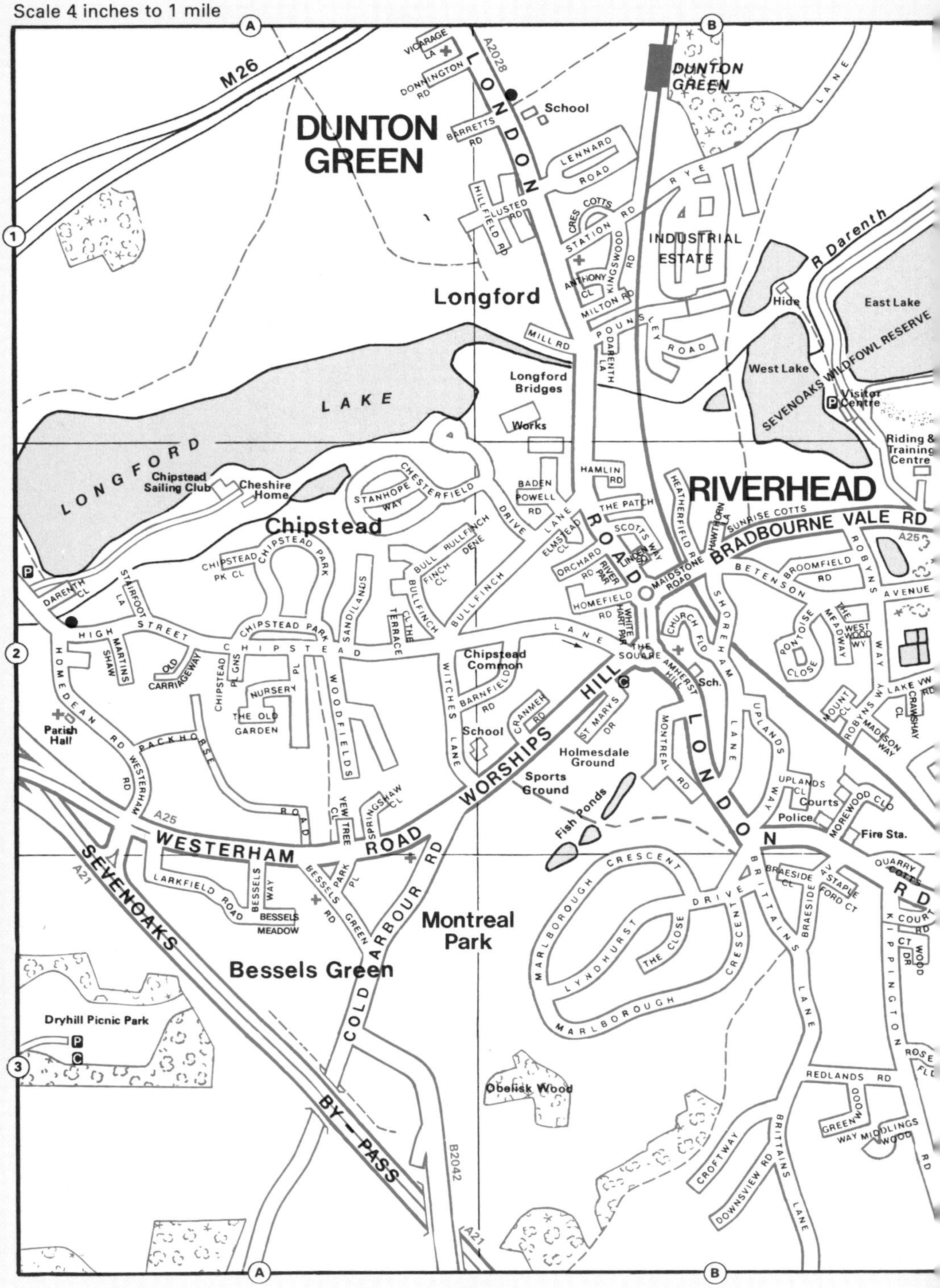

For extended coverage of this area   —   see Estate Publications RED BOOK—Sevenoaks
Reproduction prohibited without prior permission
© Estate Publications

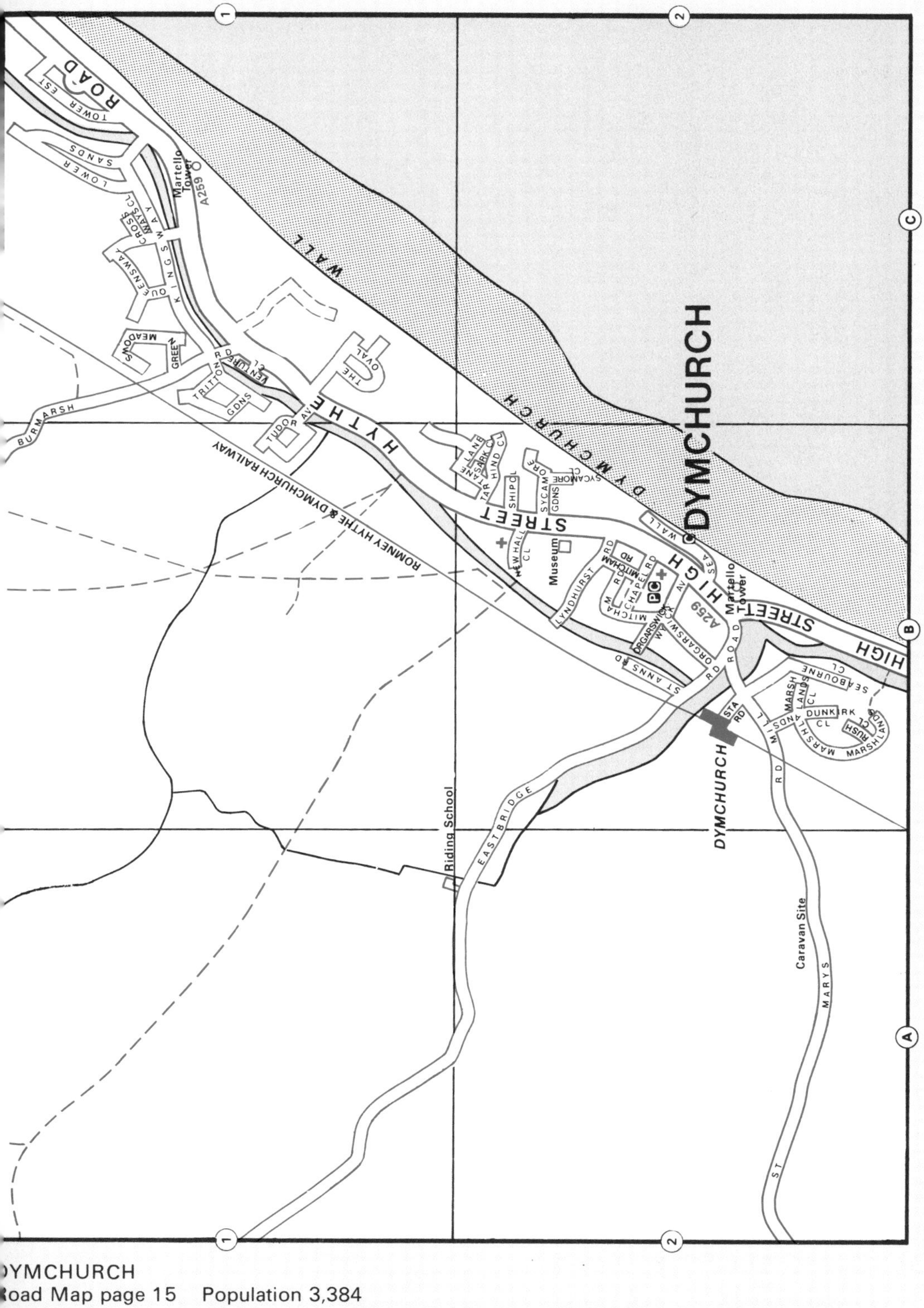

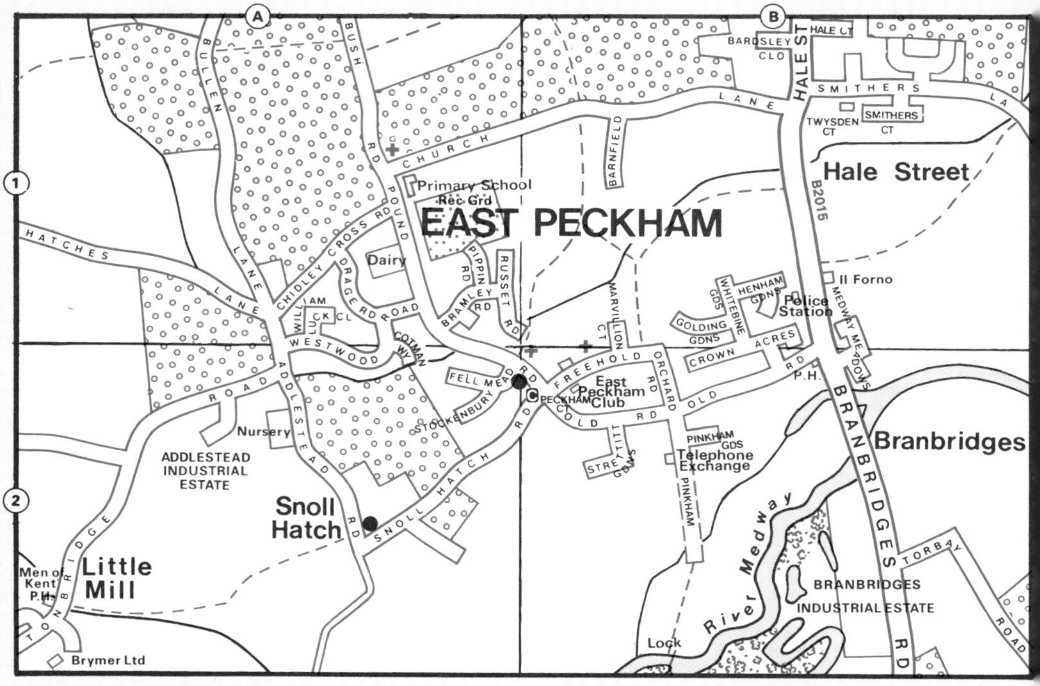

EAST PECKHAM
Road Map page 14   Population 3,287
For extended coverage of this area   —   see Estate Publications RED BOOK—Maidstone
Reproduction prohibited without prior permission   © Estate Publications

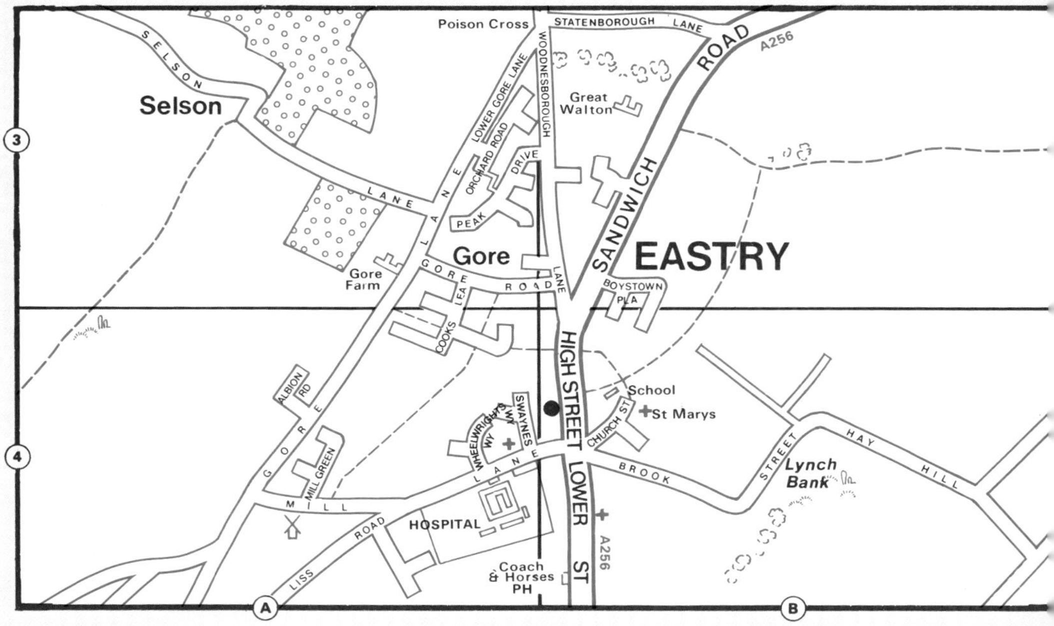

EASTRY
Road Map page 13   **EASTRY**   Scale 4 inches to 1 mi
For extended coverage of this area—see Estate Publications RED BOOK—Folkestone/Dover
Reproduction prohibited without prior permission   © Estate Publications

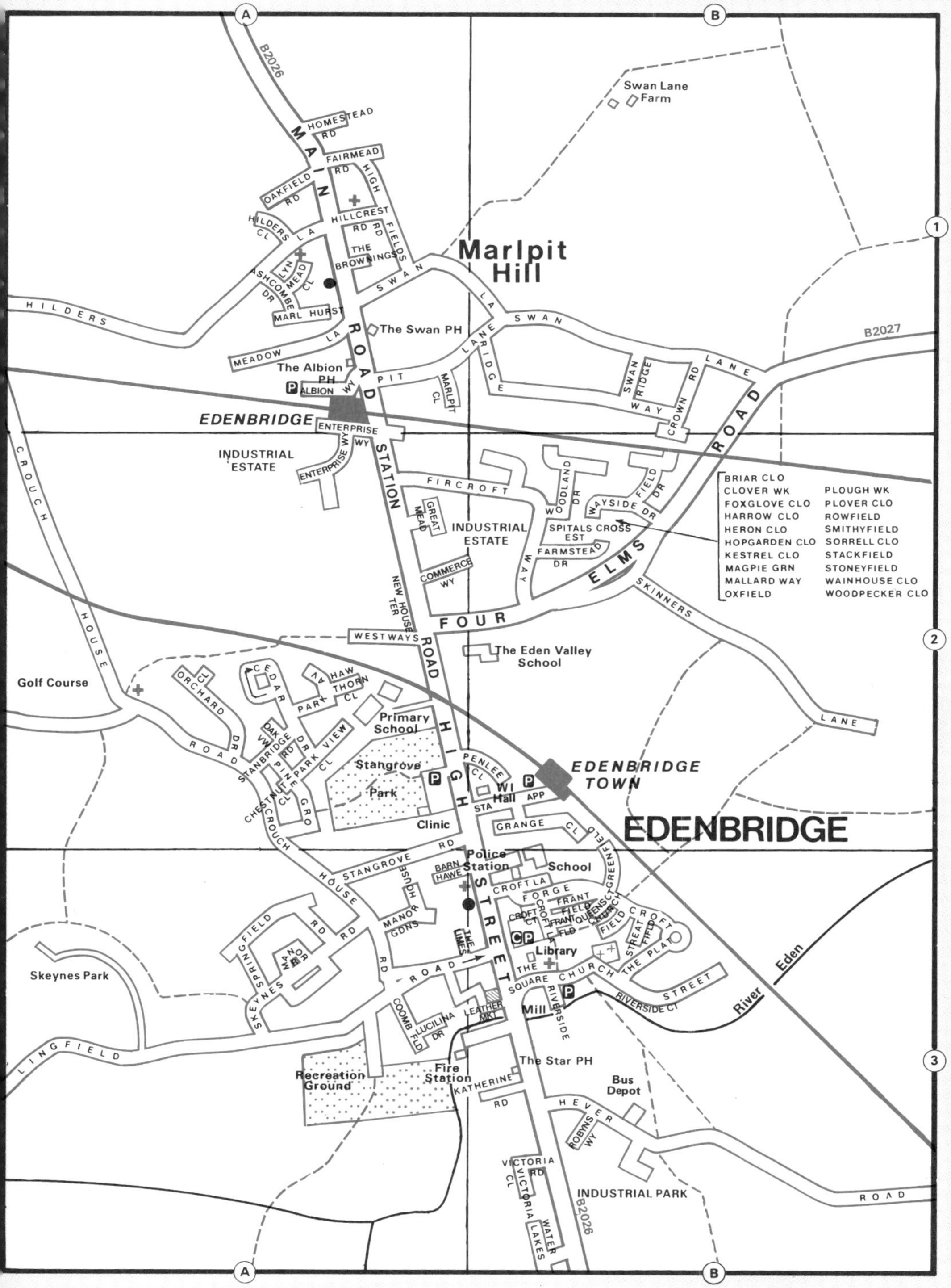

Marlpit Hill

EDENBRIDGE TOWN

EDENBRIDGE

BRIAR CLO	PLOUGH WK
CLOVER WK	PLOVER CLO
FOXGLOVE CLO	ROWFIELD
HARROW CLO	SMITHYFIELD
HERON CLO	SORRELL CLO
HOPGARDEN CLO	STACKFIELD
KESTREL CLO	STONEYFIELD
MAGPIE GRN	WAINHOUSE CLO
MALLARD WAY	WOODPECKER CLO
OXFIELD	

FAVERSHAM

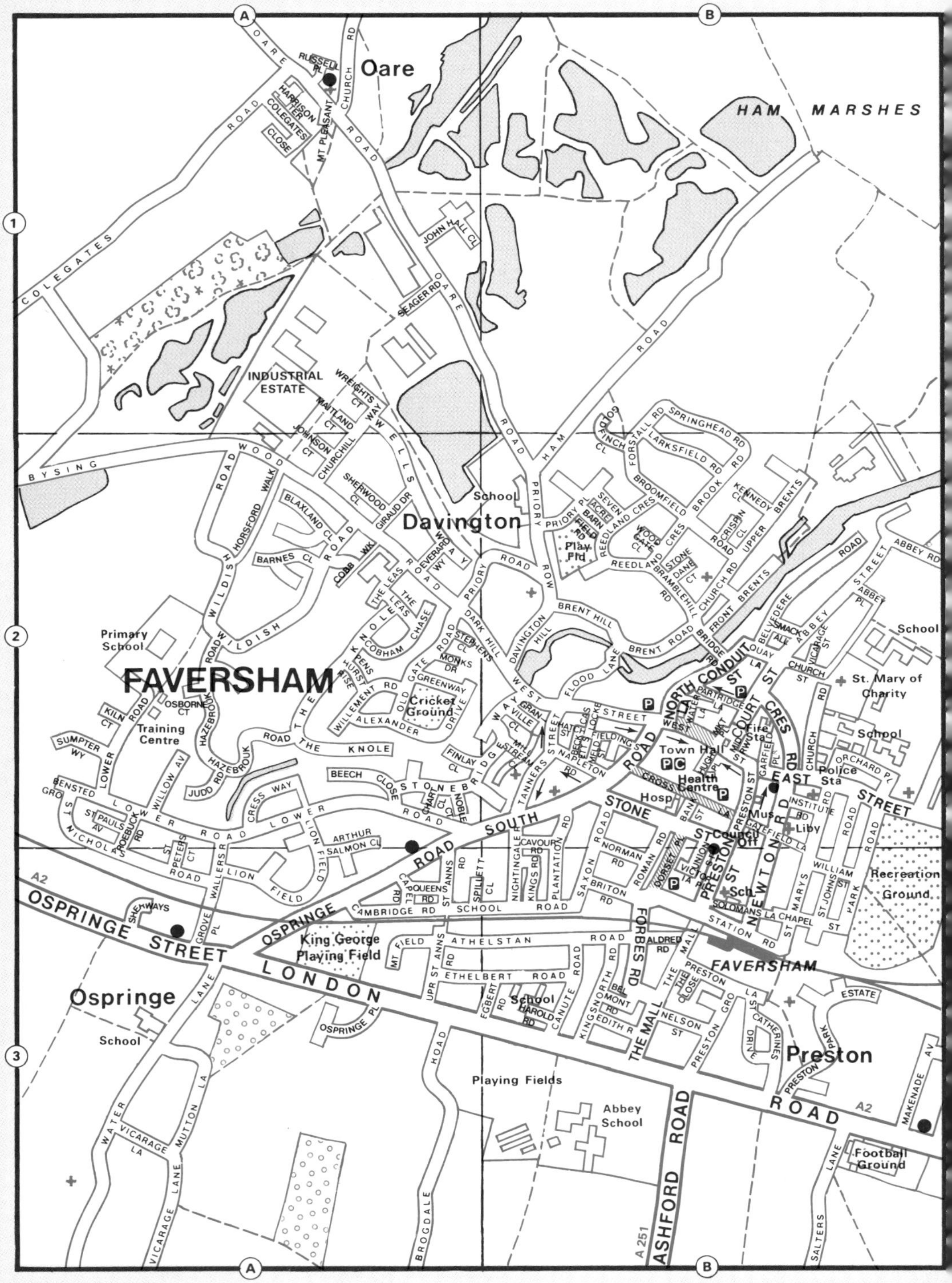

FAVERSHAM
Road Map page 12   Population 16,115
For extended coverage of this area   —   see Estate Publications RED BOOK—Swale
Reproduction prohibited without prior permission   © Estate Publications

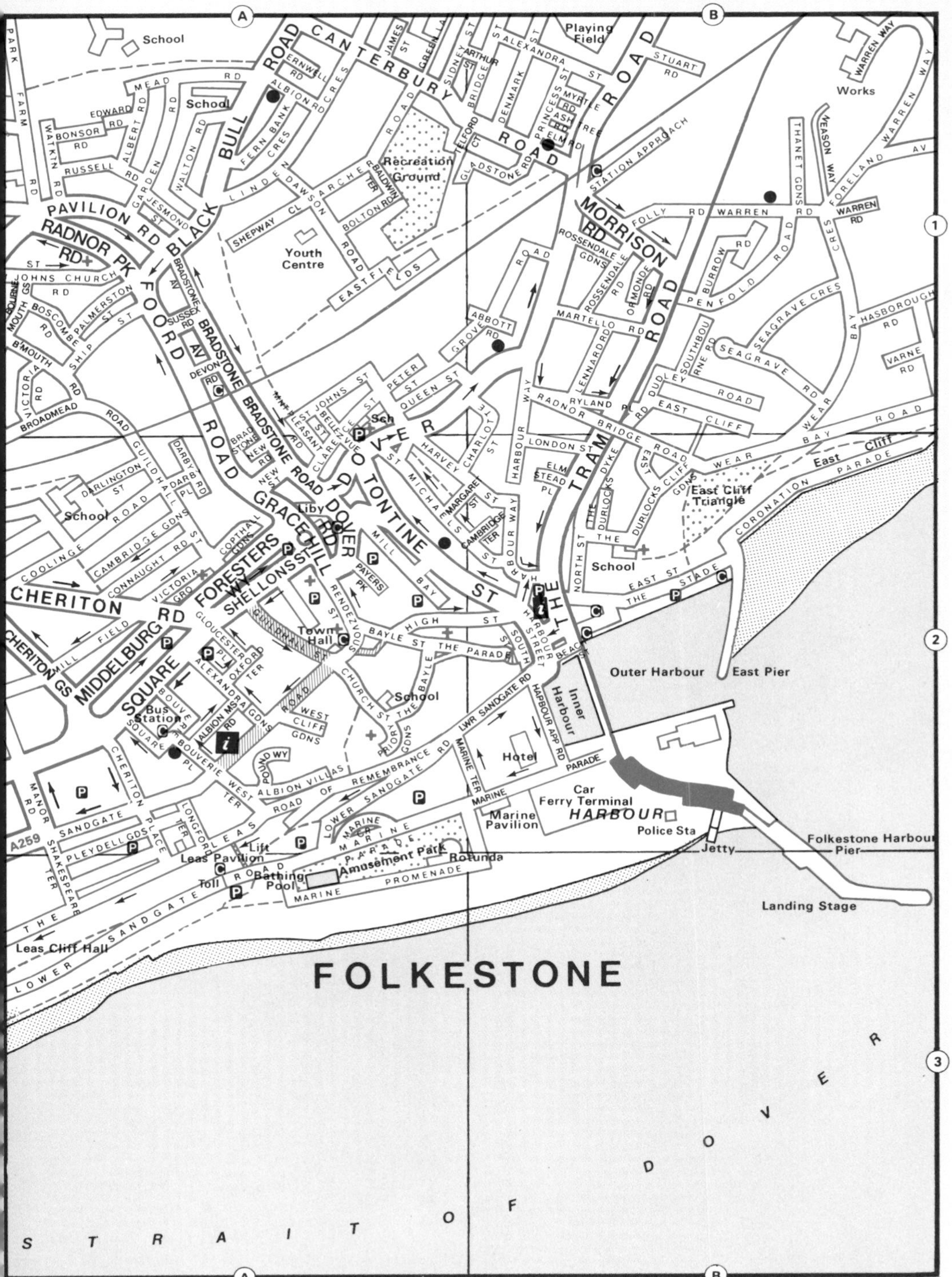

FOLKESTONE

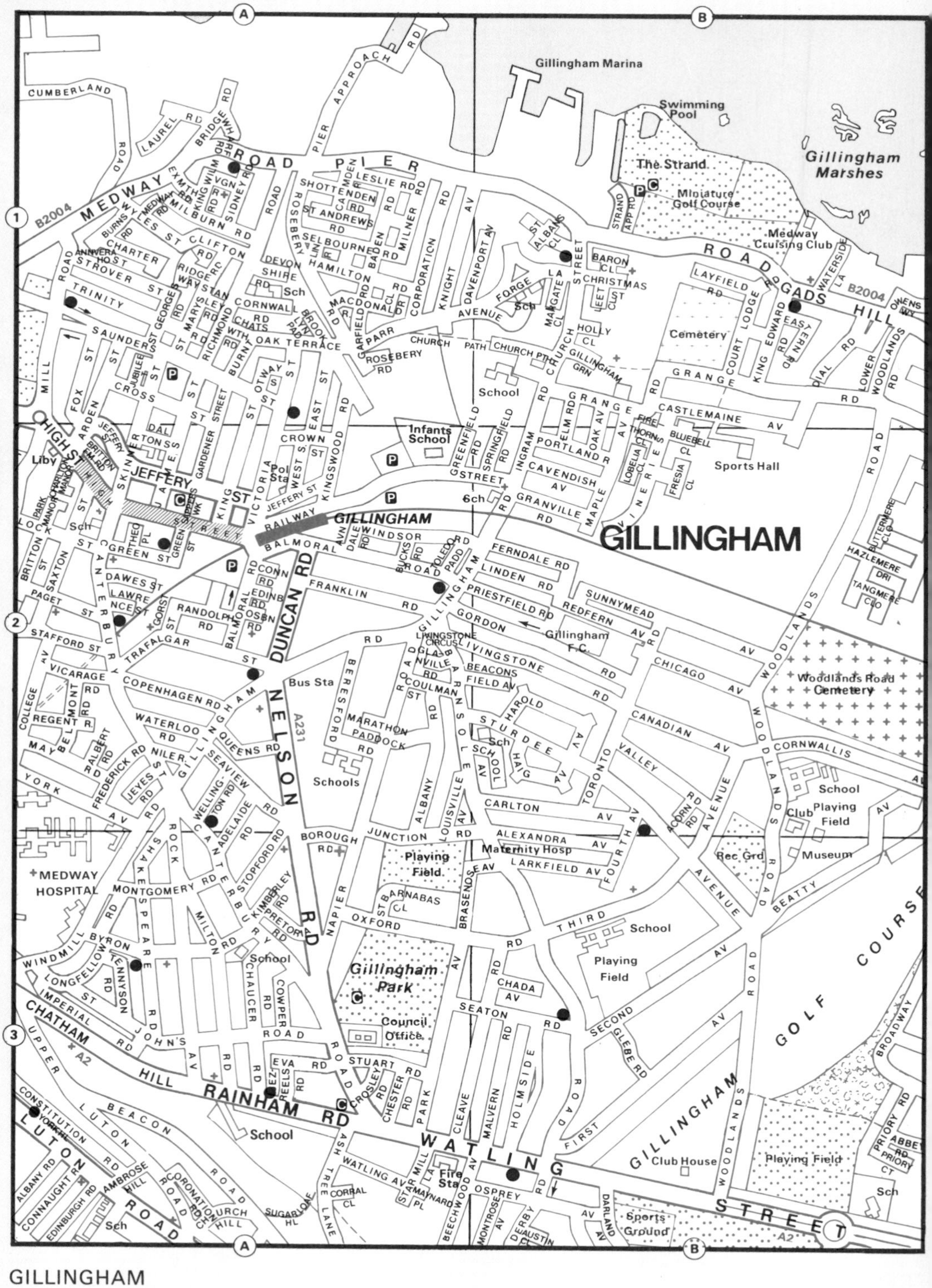

GILLINGHAM

Road Map page 11   Population 93,734

For extended coverage of this area—see Estate Publications RED BOOK—Medway/Gillingham

Reproduction prohibited without prior permission   © Estate Publications

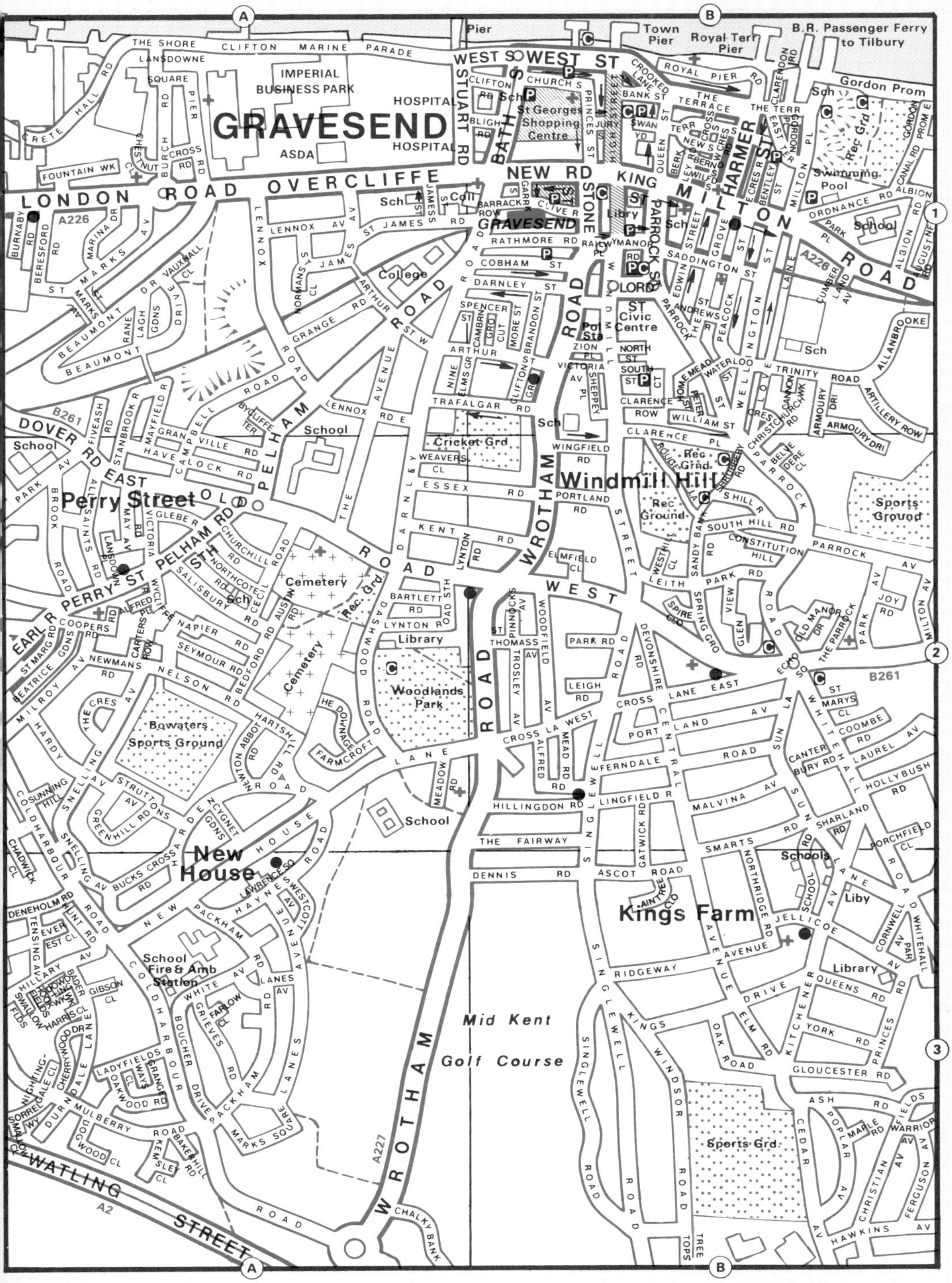

GRAVESEND
Road Map page 11   Population 52,905
For extended coverage of this area—see Estate Publications RED BOOK—Gravesend/Dartford
Reproduction prohibited without prior permission

© Estate Publications

Scale 4 inches to 1 mile

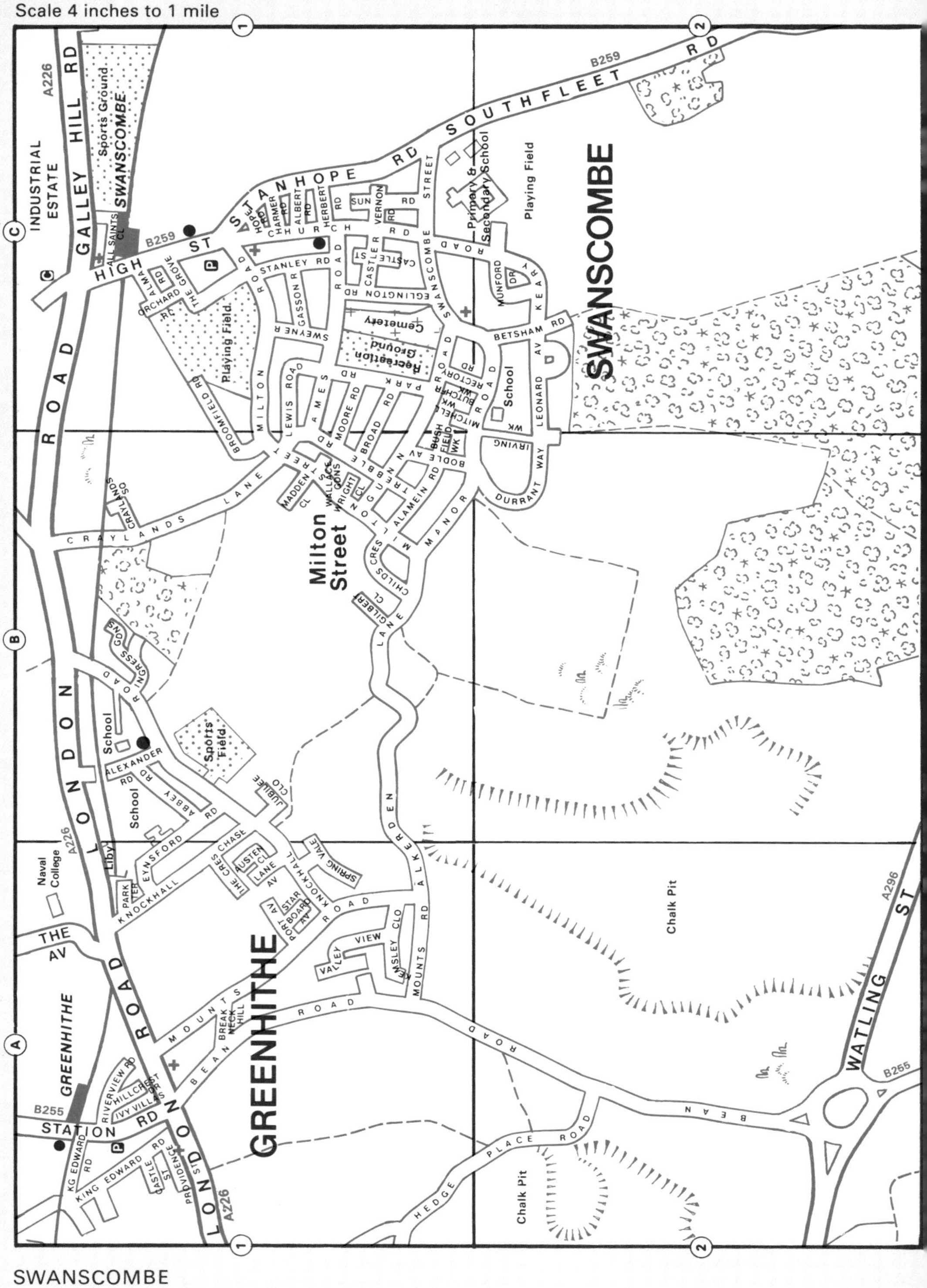

SWANSCOMBE
Road Map page 10   Population 8,876
For extended coverage of this area—see Estate Publications RED BOOK—Gravesend/Dartford
Reproduction prohibited without prior permission   © Estate Publications

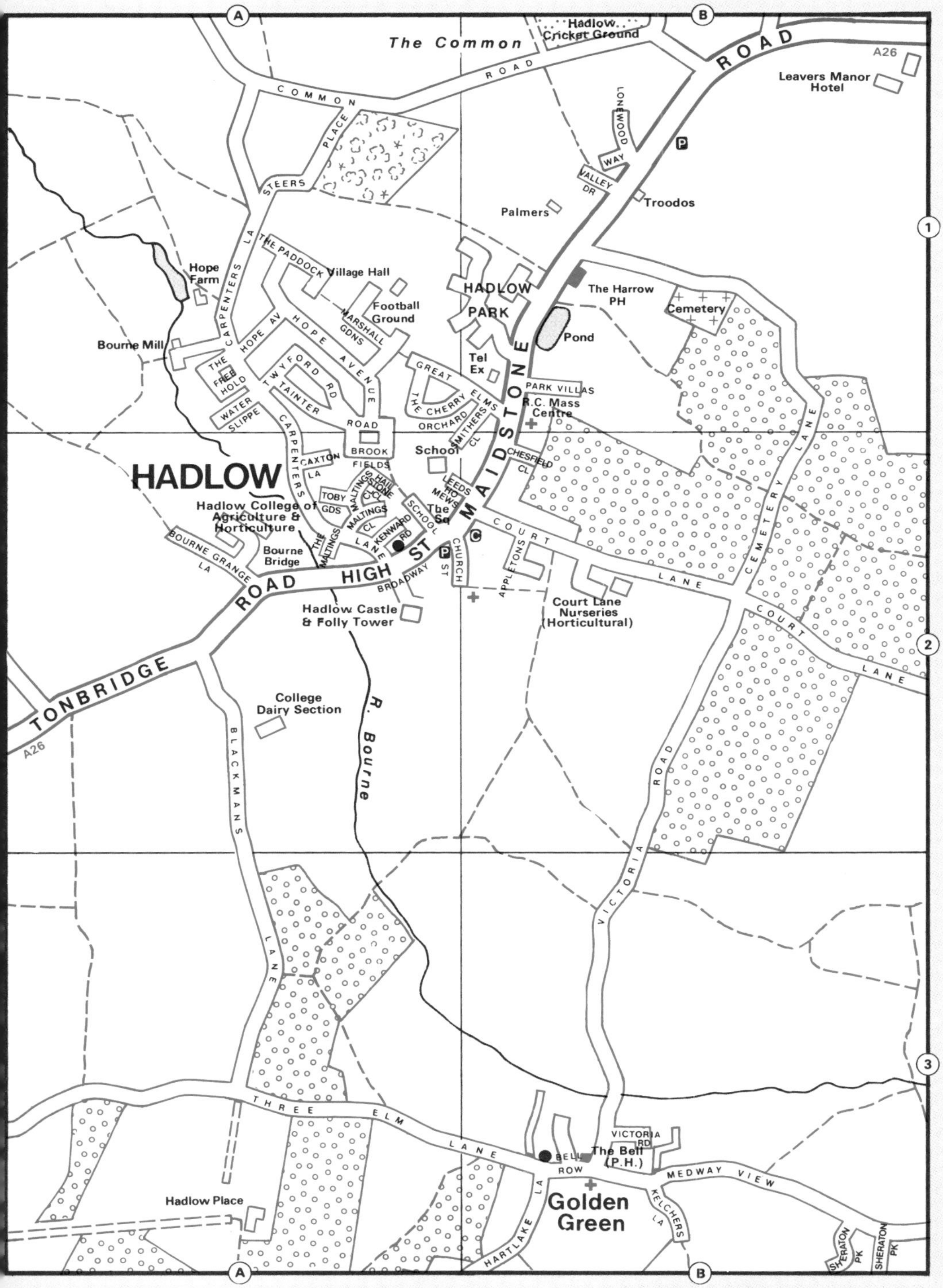

HADLOW
Road Map page 14   Population 3,365
For extended coverage of this area—

Reproduction prohibited without prior permission
© Estate Publications

see Estate Publications RED BOOK—Tunbridge Wells/Tonbridge

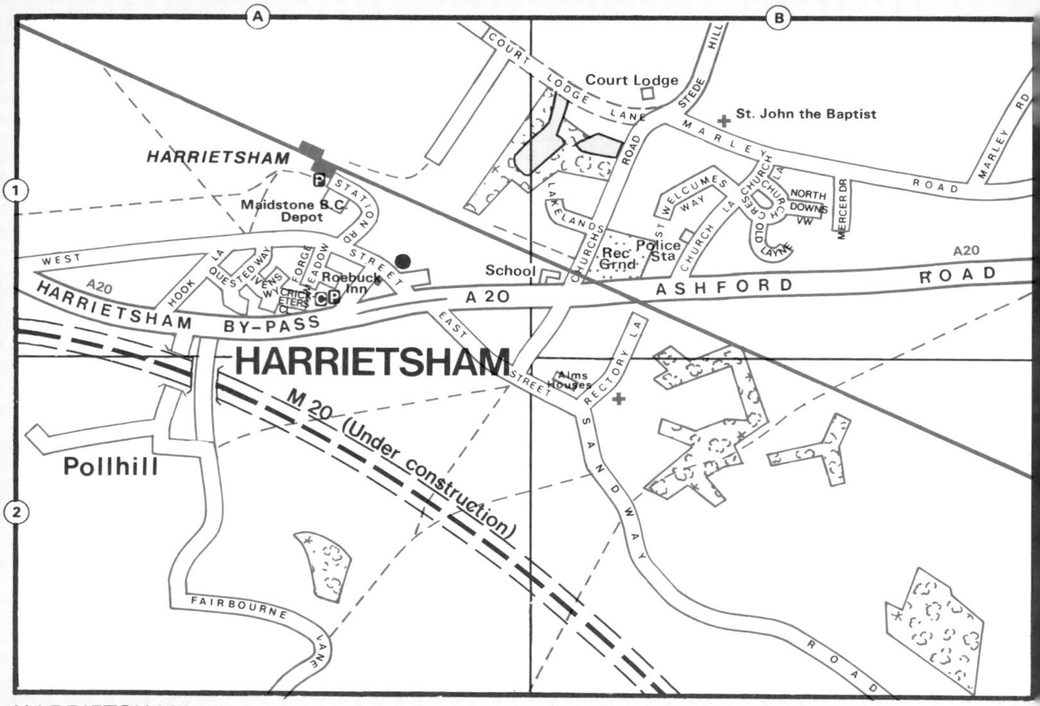

HARRIETSHAM

Road Map page 15   Population 1,404

Fo extended coverage of this area   —   see Estate Publications RED BOOK—Maidstone

Reproduction prohibited without prior permission   © Estate Publications

Scale 4 inches to 1 mile

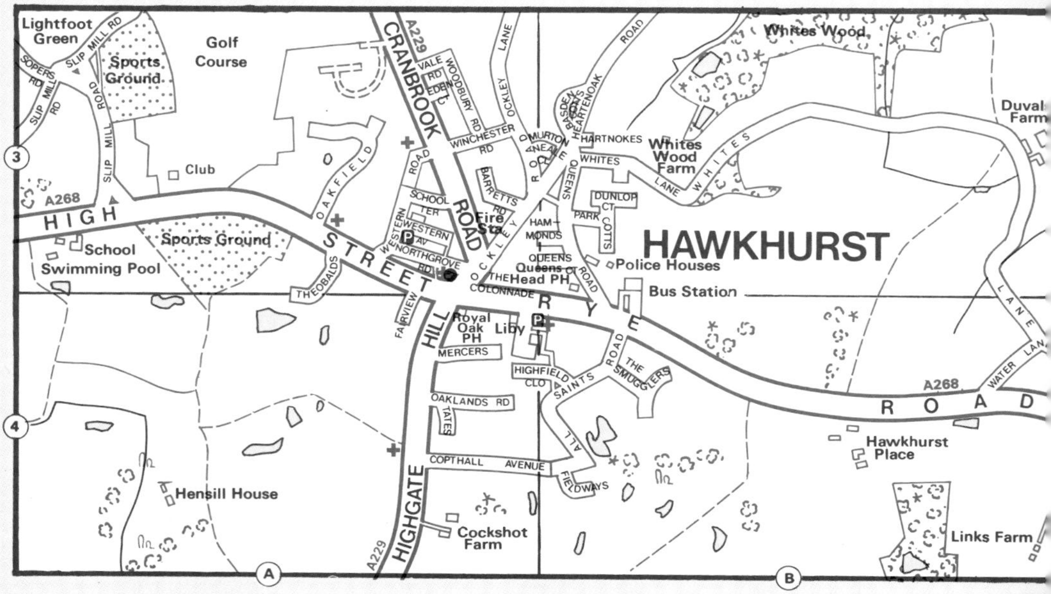

HAWKHURST

HAWKHURST   **HAWKHURST**   Population 3,968

Road Map page 14

Reproduction prohibited without prior permission   © Estate Publications

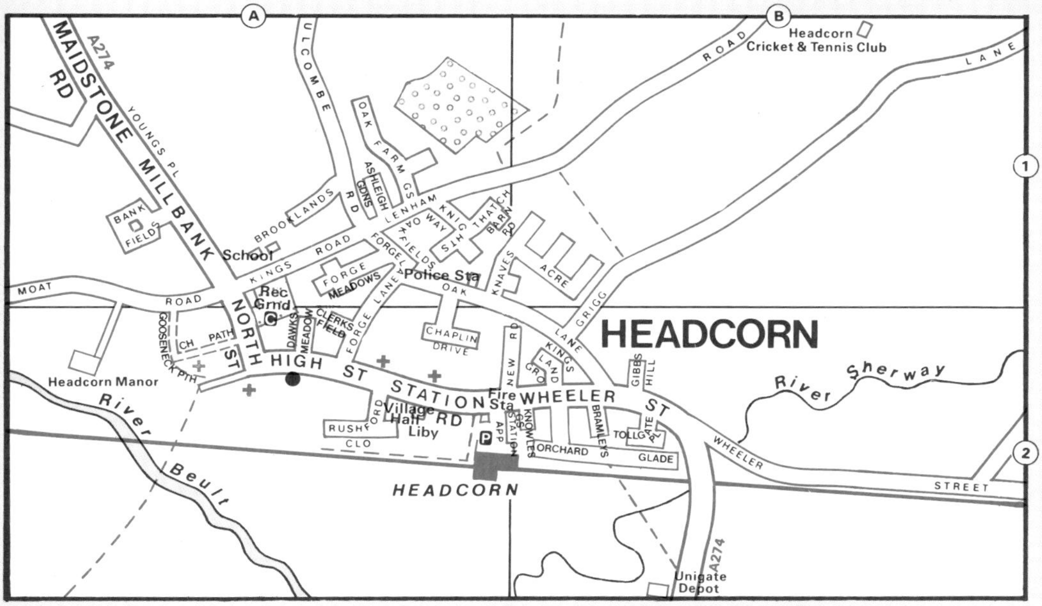

HEADCORN

HEADCORN
Road Map page 15   Population 3,208

For extended coverage of this area —
see Estate Publications RED BOOK—Maidstone

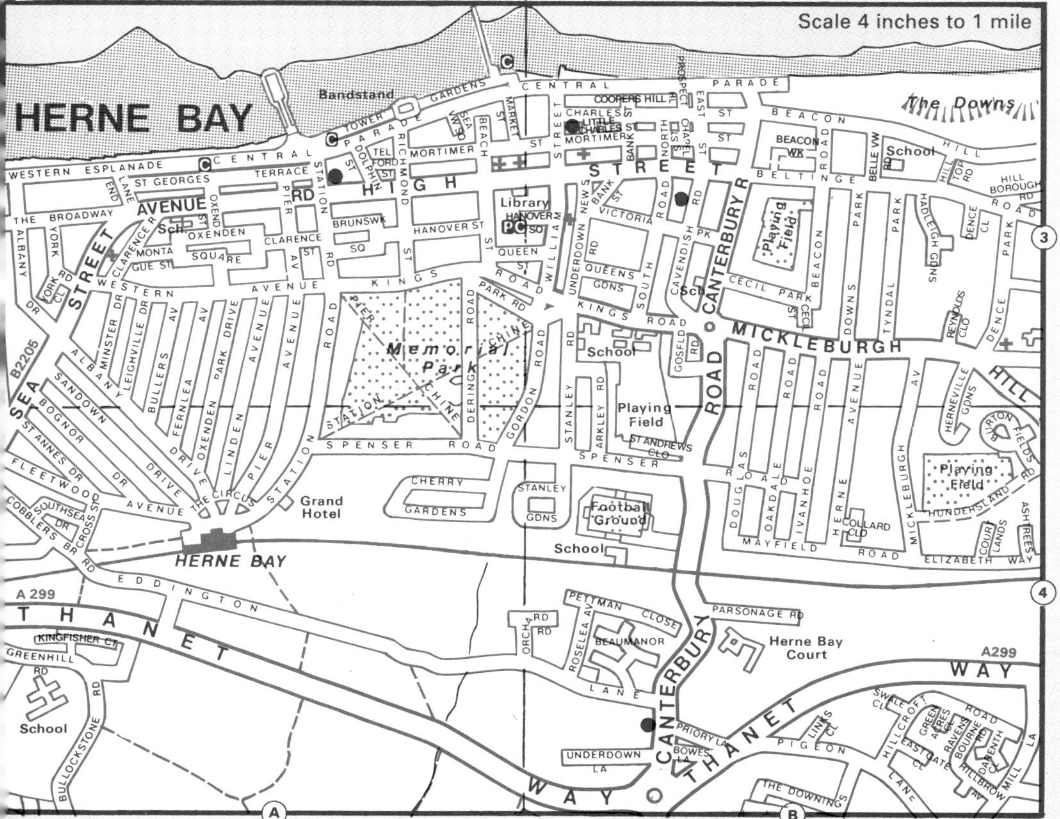

HERNE BAY

Population 28,143

HERNE BAY
Road Map page 13

For extended coverage of this area—see Estate Publications RED BOOK—Thanet/Canterbury

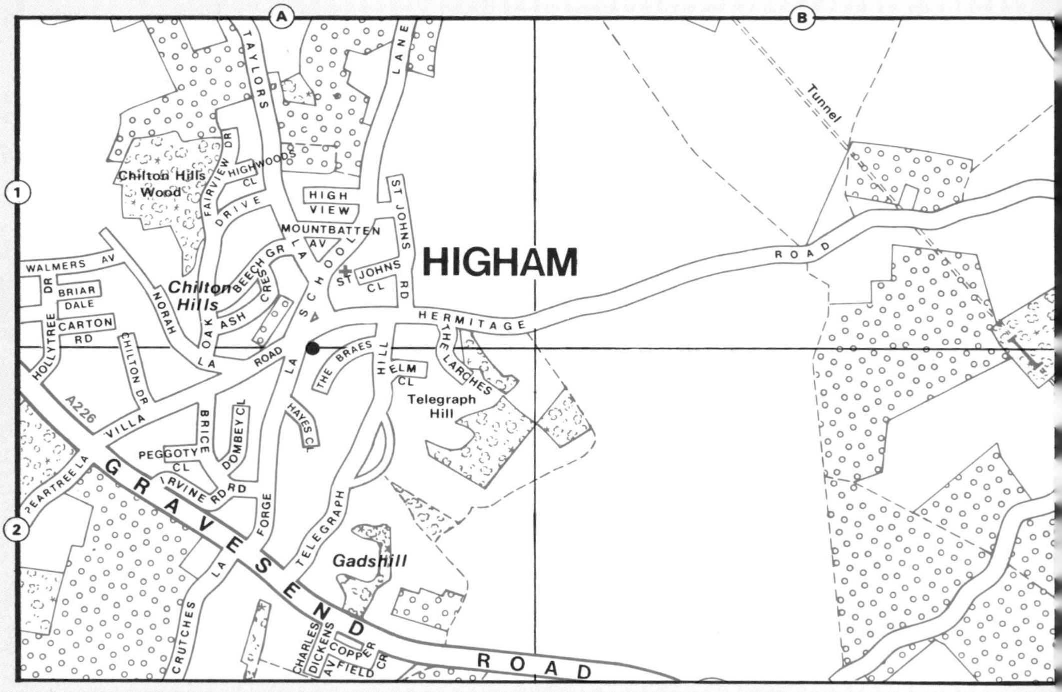

HIGHAM   Population 4,000
Road Map page 11

For extended coverage of this area —
see Estate Publications RED BOOK—Medway/Gillingham

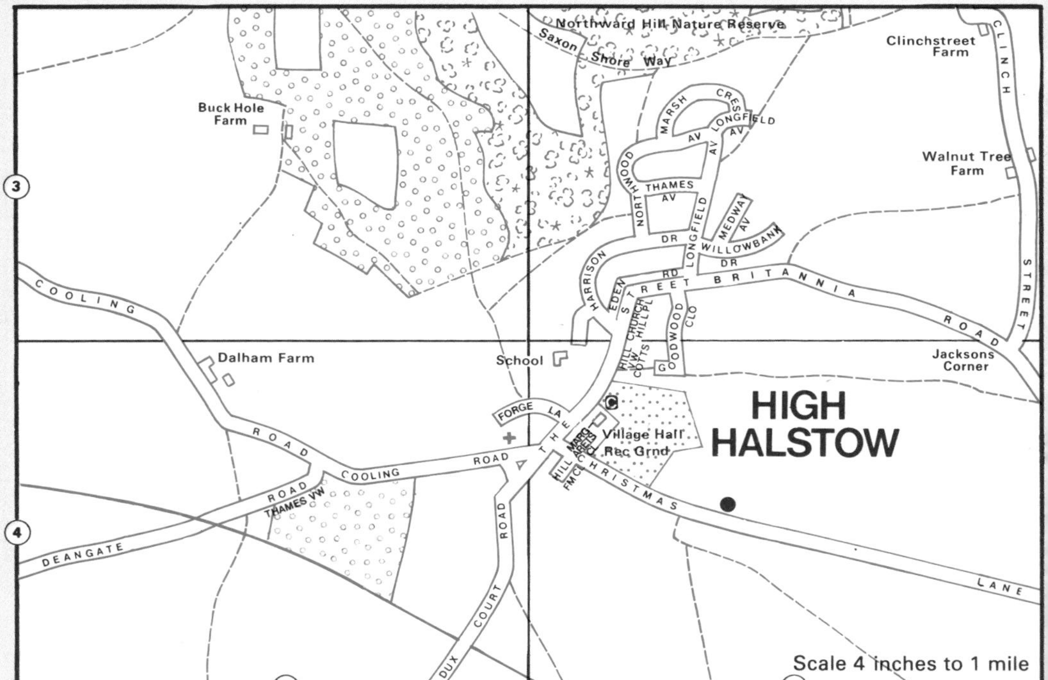

HIGH HALSTOW
Road Map page 11

HIGH HALSTOW

Population 1,331

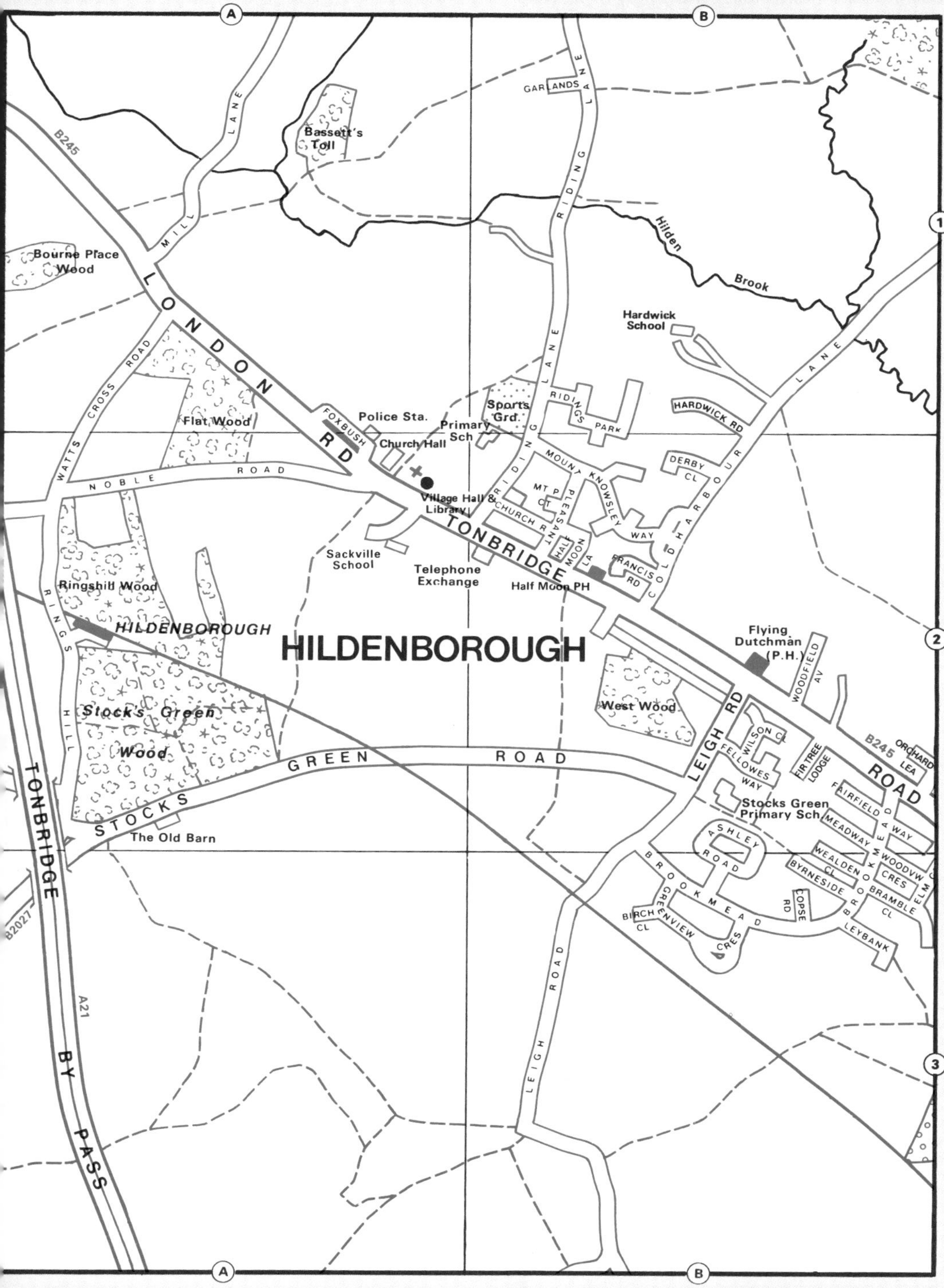

HILDENBOROUGH

For extended coverage of this area— see Estate Publications RED BOOK—Tunbridge Wells
/Tonbridge

© Estate Publications

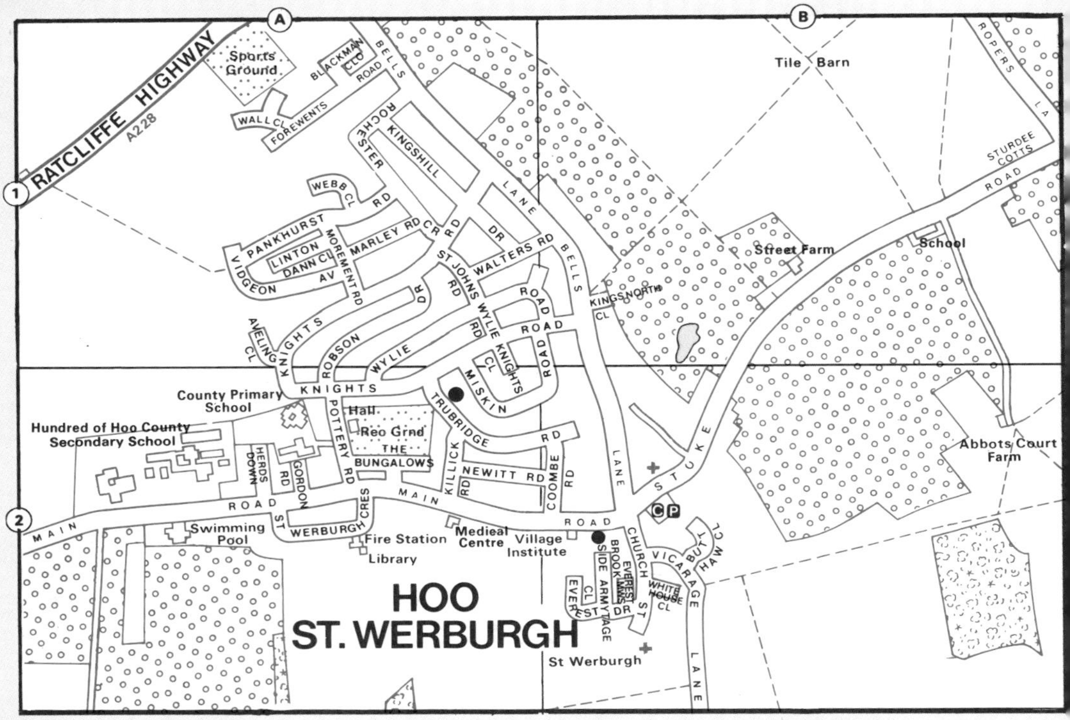

HOO ST. WERBURGH
Road Map page 11   Population 8,105

For extended coverage of this area — see
Estate Publications RED BOOK—Medway/Gillingham

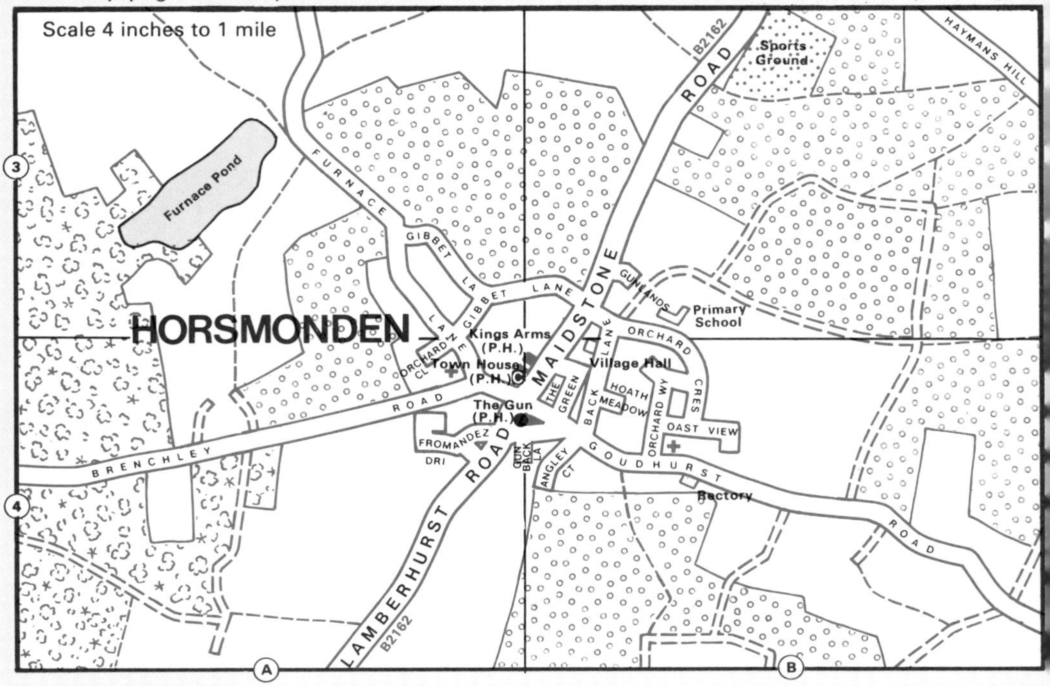

HORSMONDEN
Road Map page 14

HORSMONDEN

Population 1,907

For extended coverage of this area — see Estate Publications RED BOOK—Tunbridge Wells/
Tonbridge

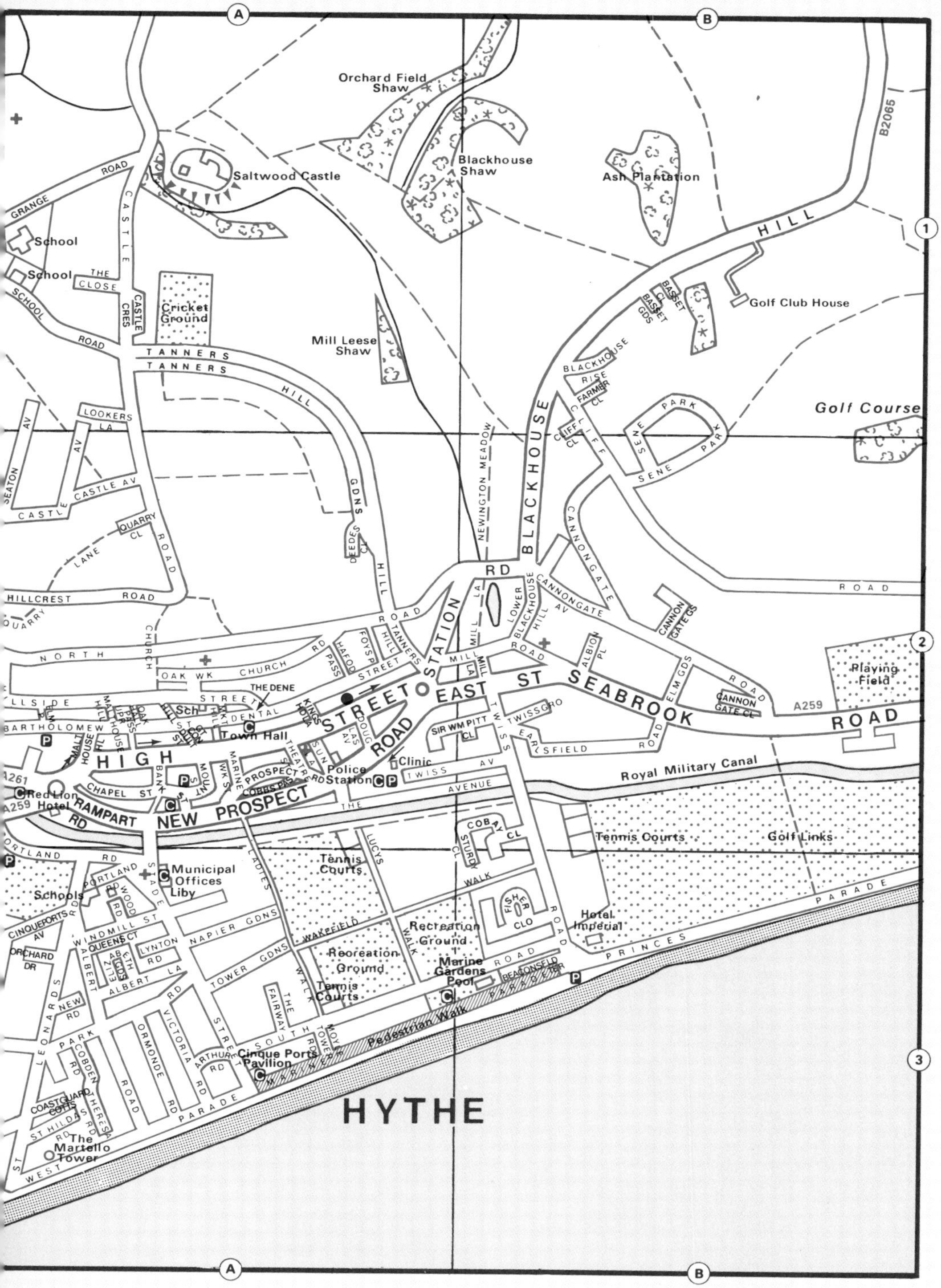

HYTHE

YTHE
oad Map page 13   Population 12,823
or extended coverage of this area—see Estate Publications RED BOOK—Folkestone/Dover

IGHTHAM

IGHTHAM
Road Map page 10   Population 1,731
For extended coverage of this area   —   see Estate Publications RED BOOK—Sevenoaks

Scale 4 inches to 1 mile

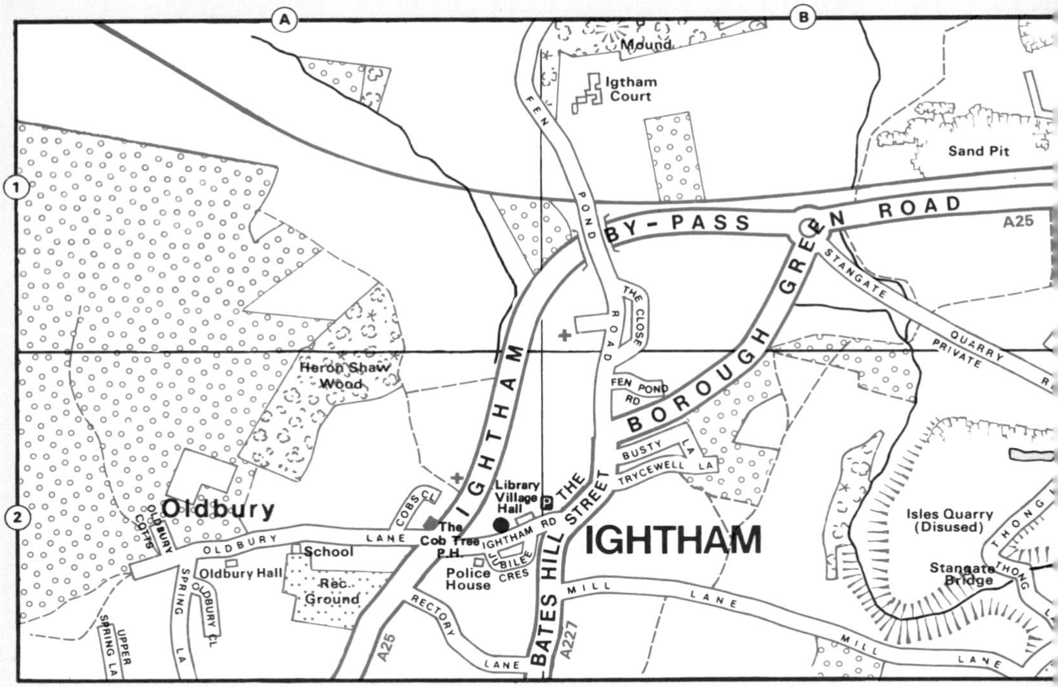

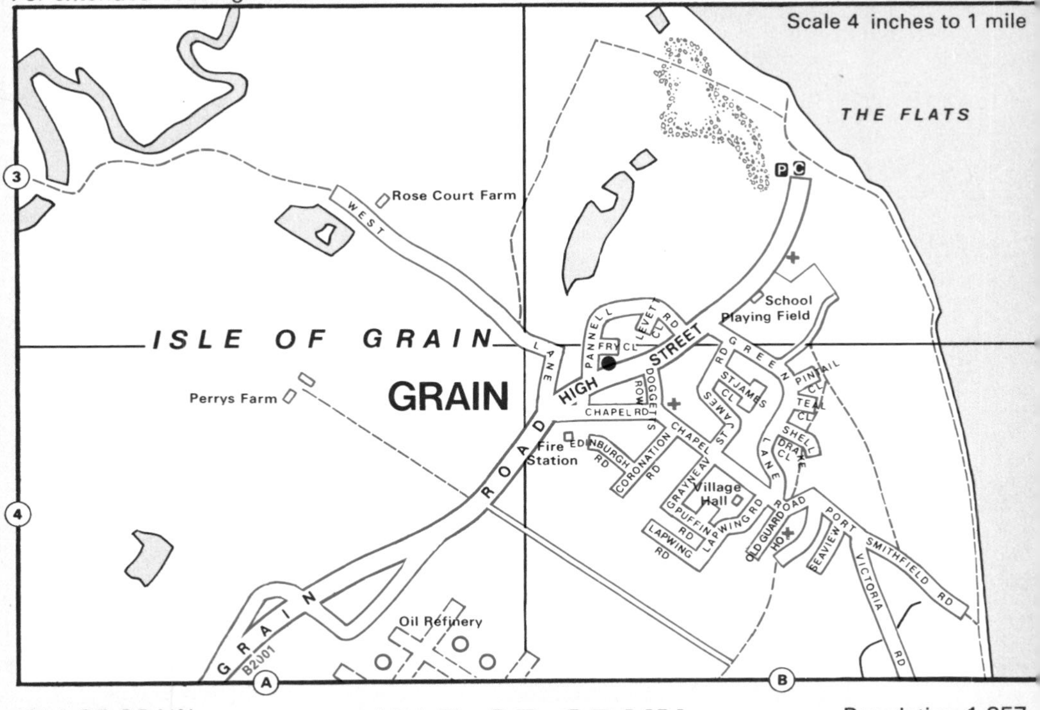

ISLE OF GRAIN

GRAIN

THE FLATS

ISLE OF GRAIN
Road Map page 11

ISLE OF GRAIN

Population 1,857

For extended coverage of this area—see Estate Publications RED BOOK—Medway/Gillingham

KEMSING

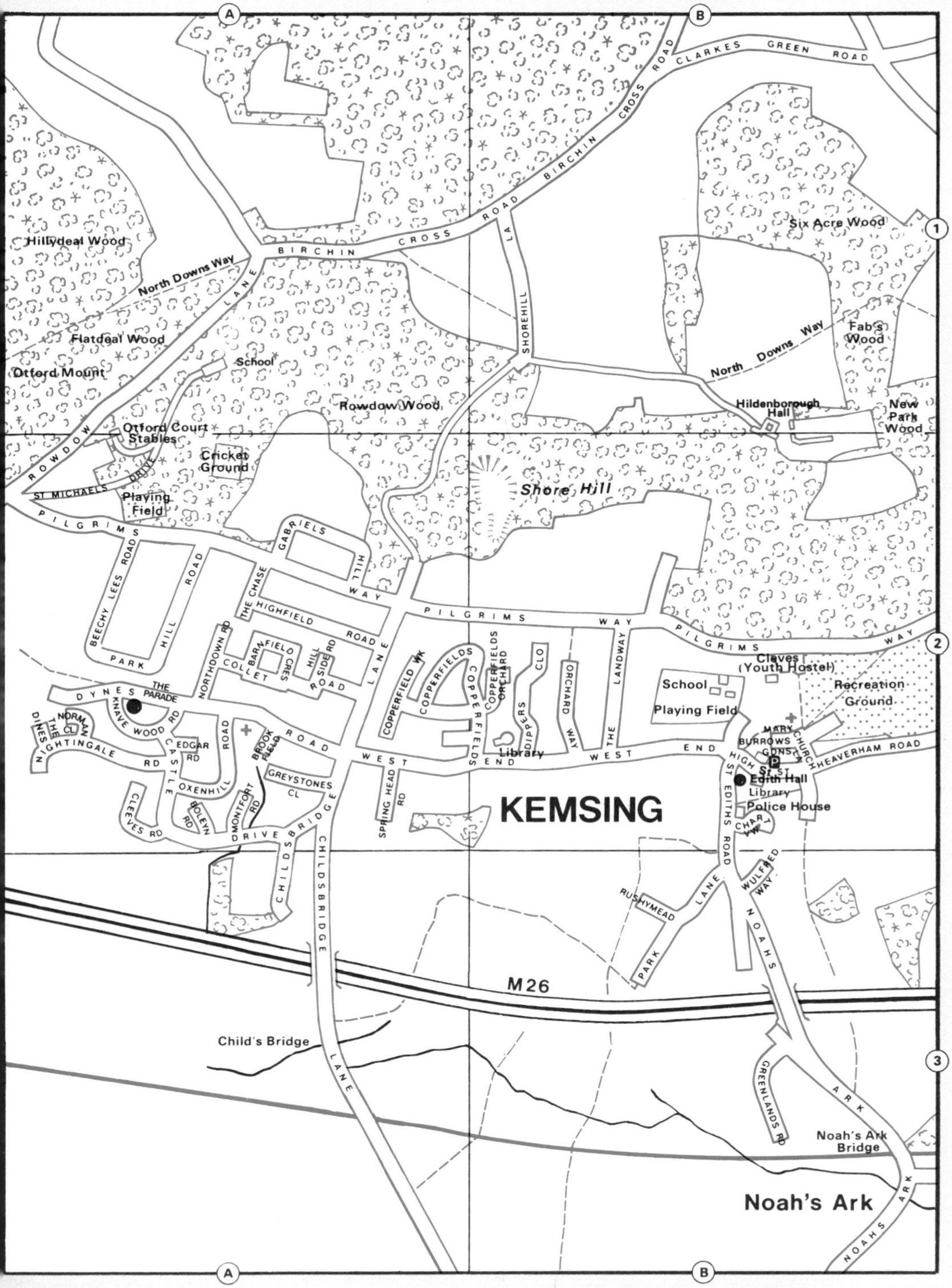

KINGSDOWN

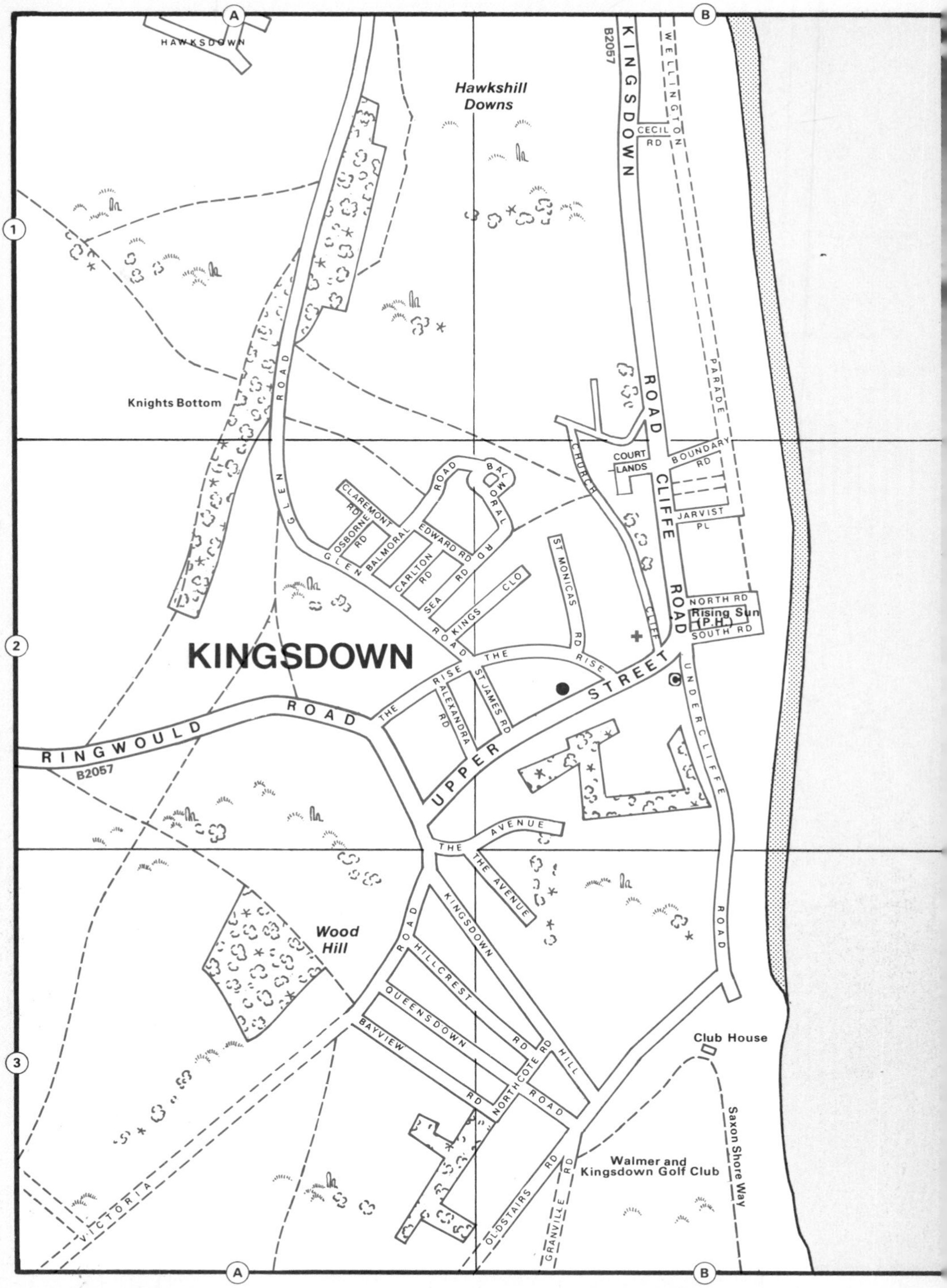

For extended coverage of this area—see Estate Publications RED BOOK—Folkestone/Dover

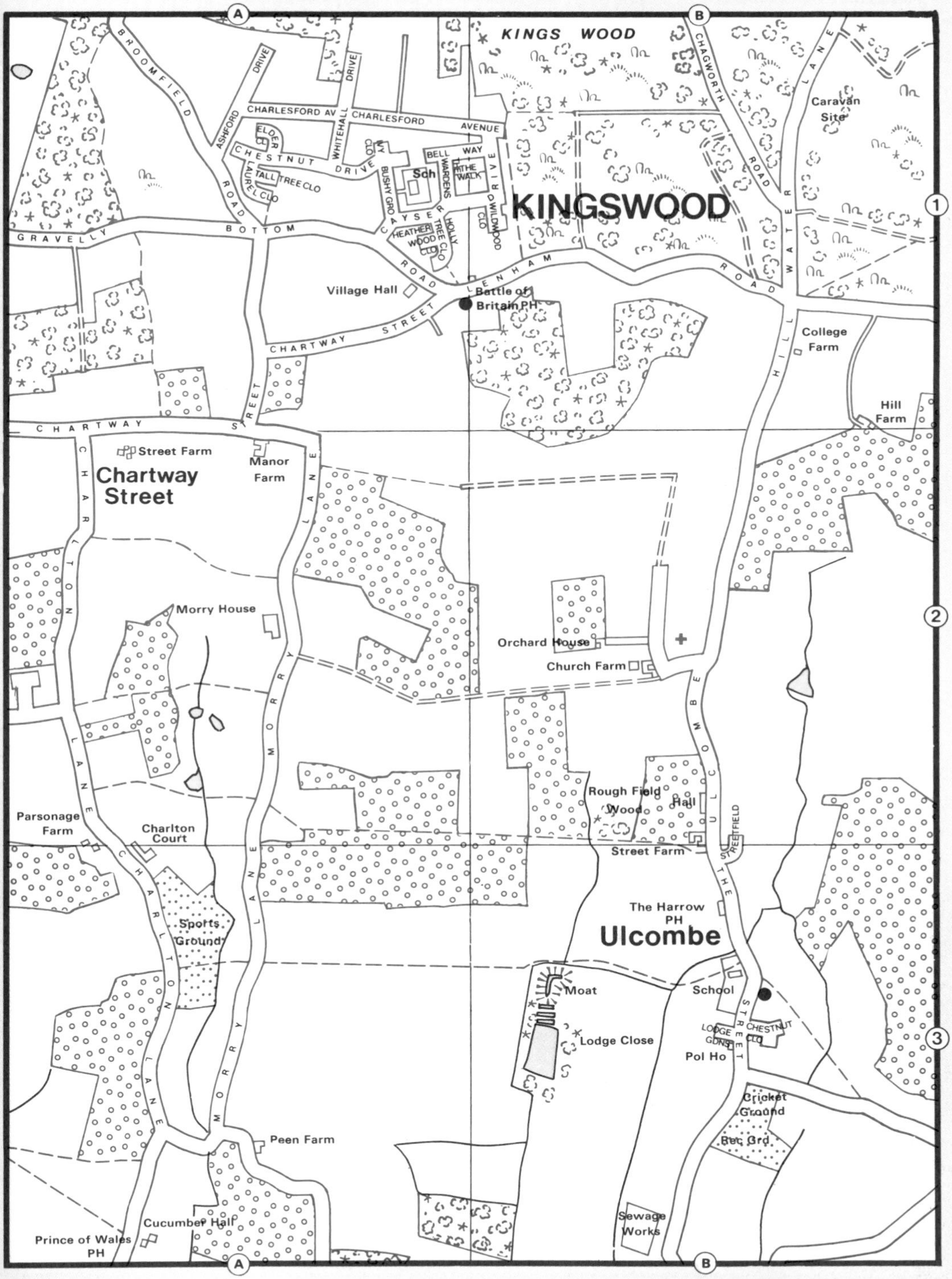

KINGSWOOD
Road Map page 15
For extended coverage of this area — see Estate Publications RED BOOK—Maidstone
Reproduction prohibited without prior permission

ULCOMBE Population 836

LANGLEY HEATH

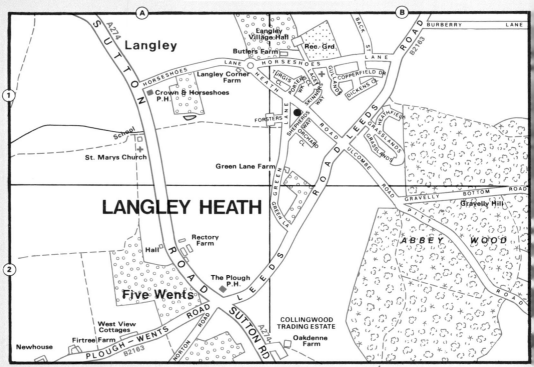

LANGLEY HEATH
Road Map page 11

Scale 4 inches to 1 mile

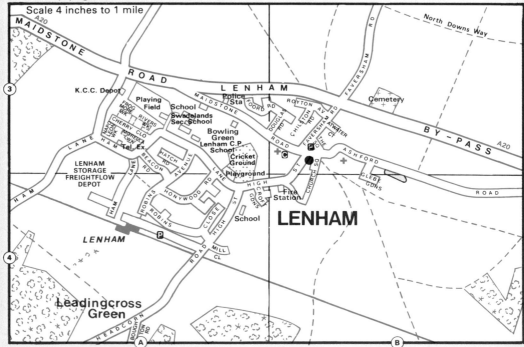

LENHAM

Population 3,549

For extended coverage of this area — see Estate Publications RED BOOK—Maidstone

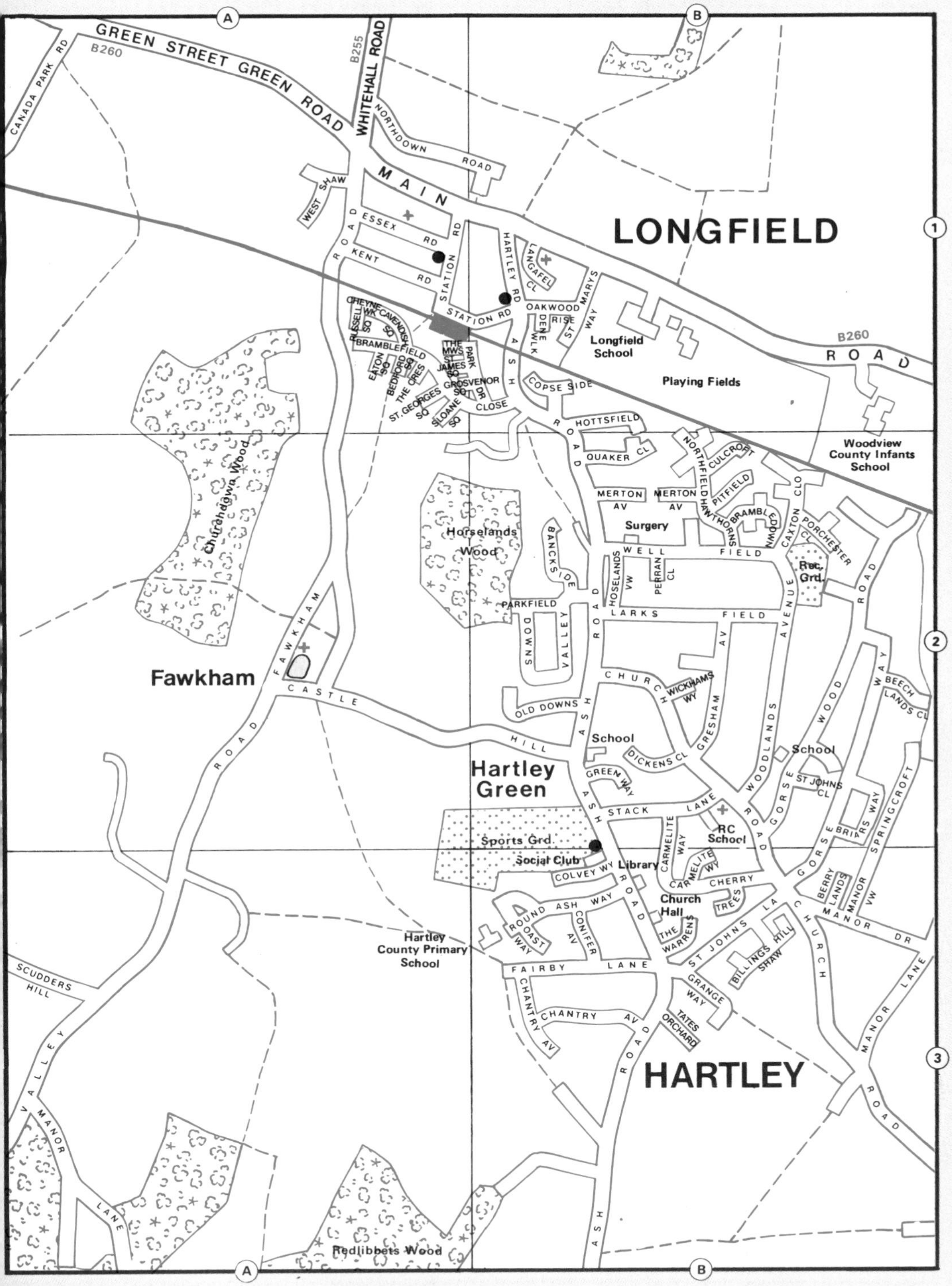

For extended coverage of this area—see Estate Publications RED BOOK—Gravesend/Dartford

LYDD
Road Map page 15   Population 4,729

For extended coverage of this area — see
Estate Publications RED BOOK — Rye & Romney Marsh

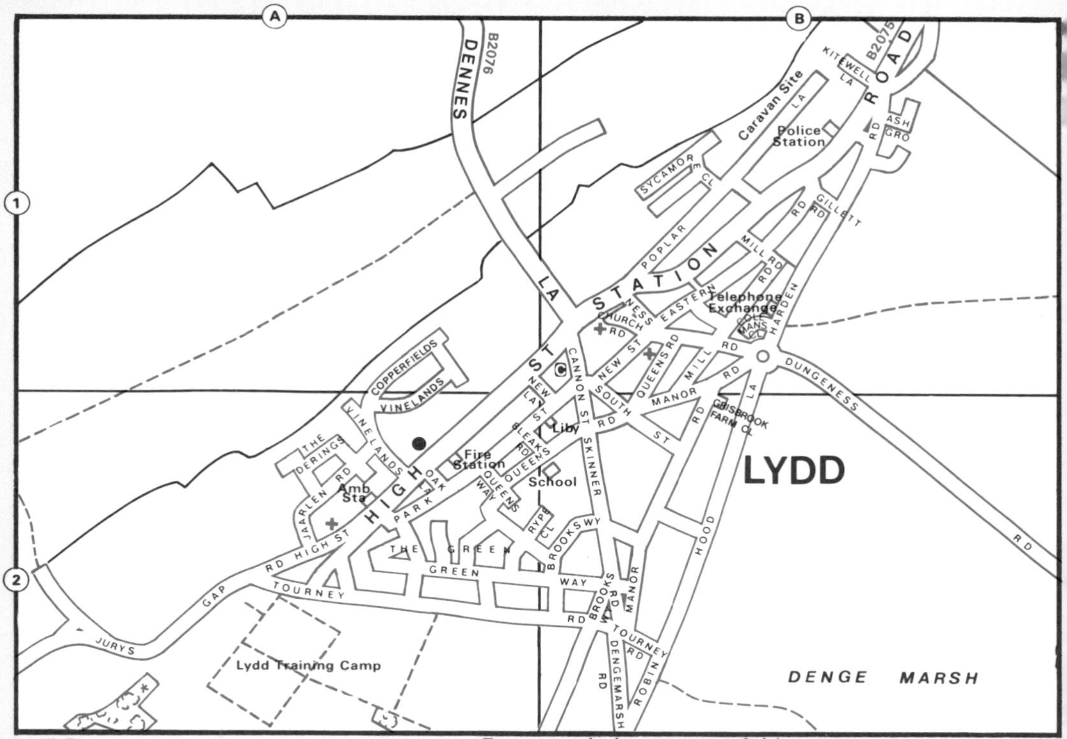

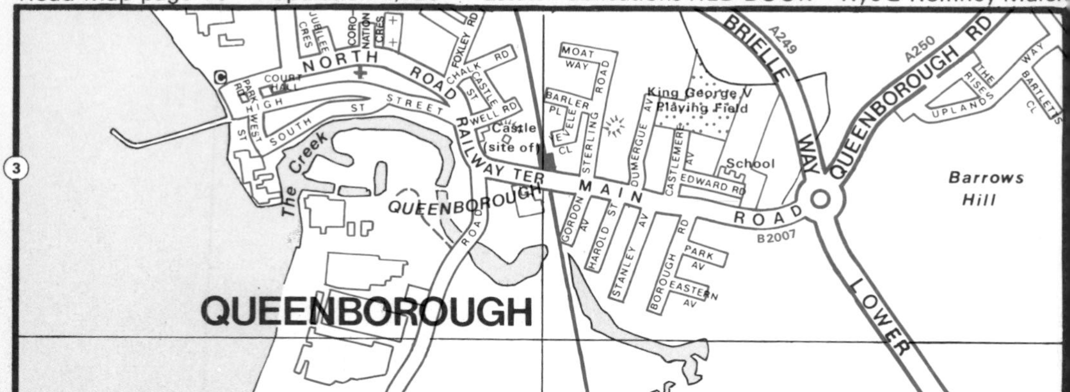

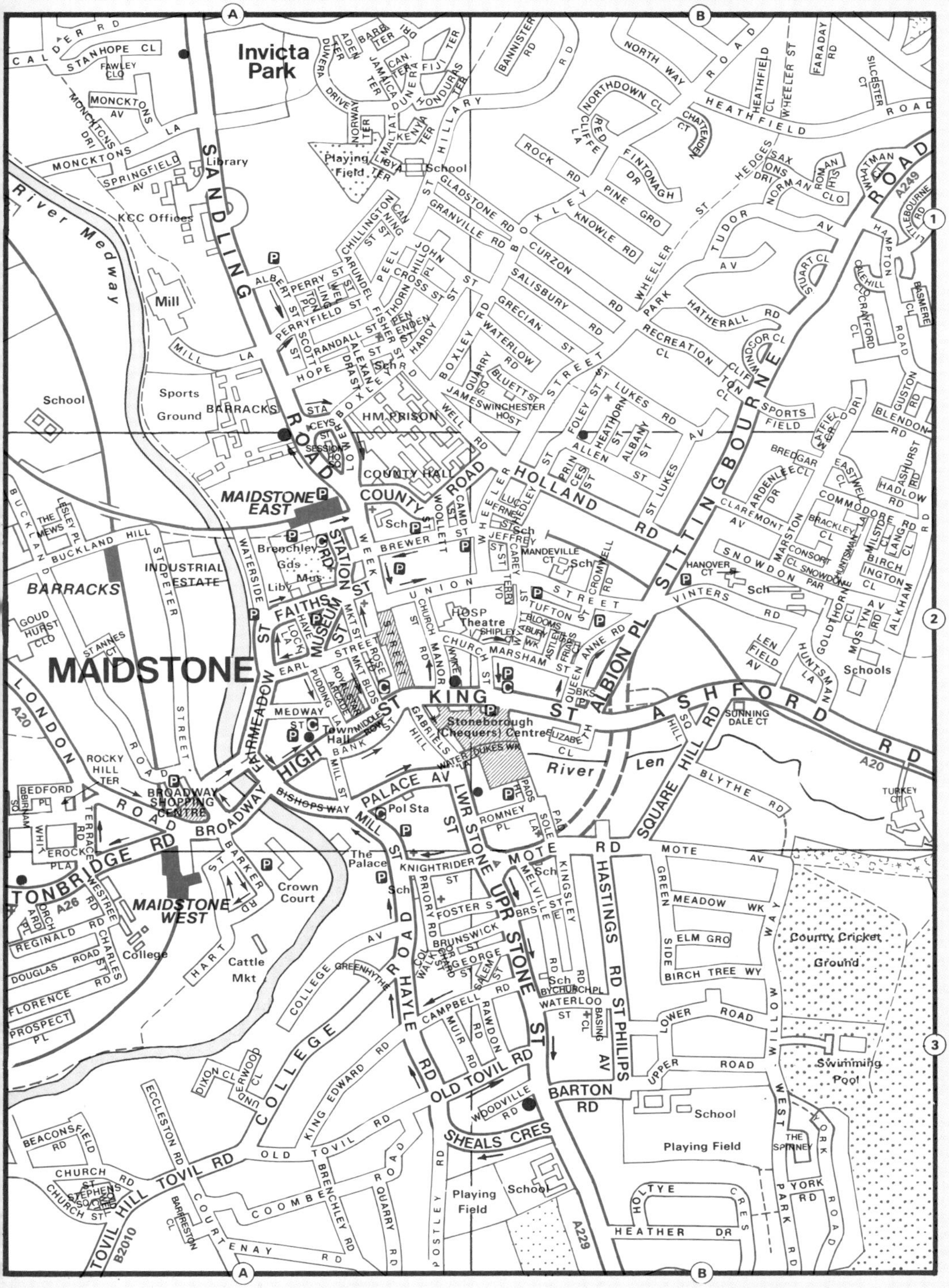

MAIDSTONE
Road Map page 11   Population 72,494
For extended coverage of this area — see Estate Publications RED BOOK—Maidstone

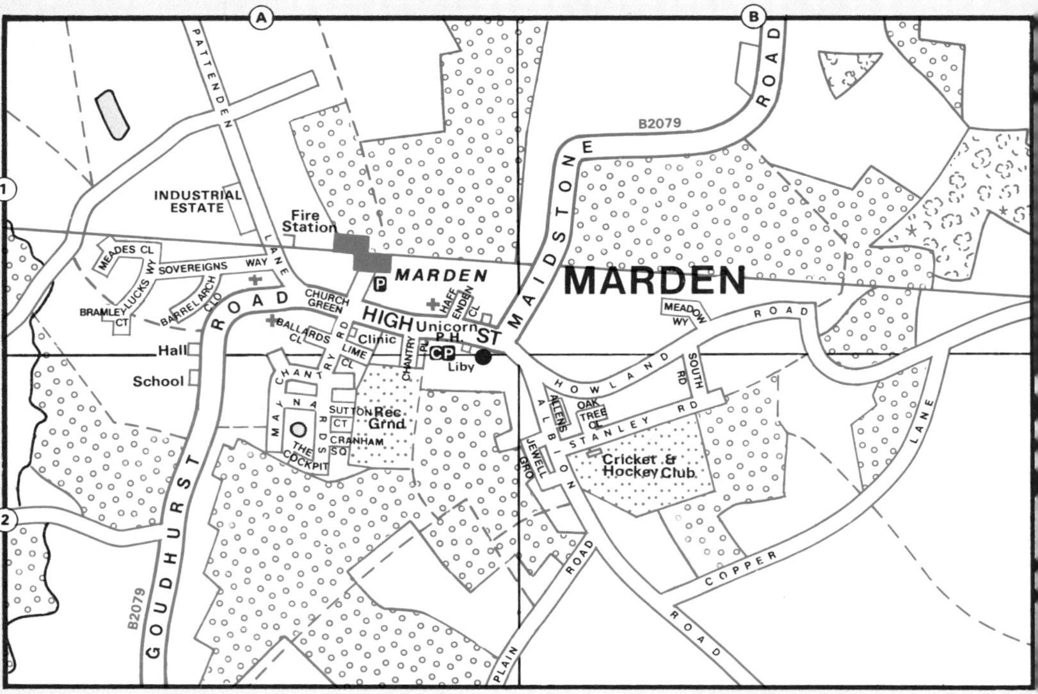

MARDEN
Road Map page 14   Population 3,294

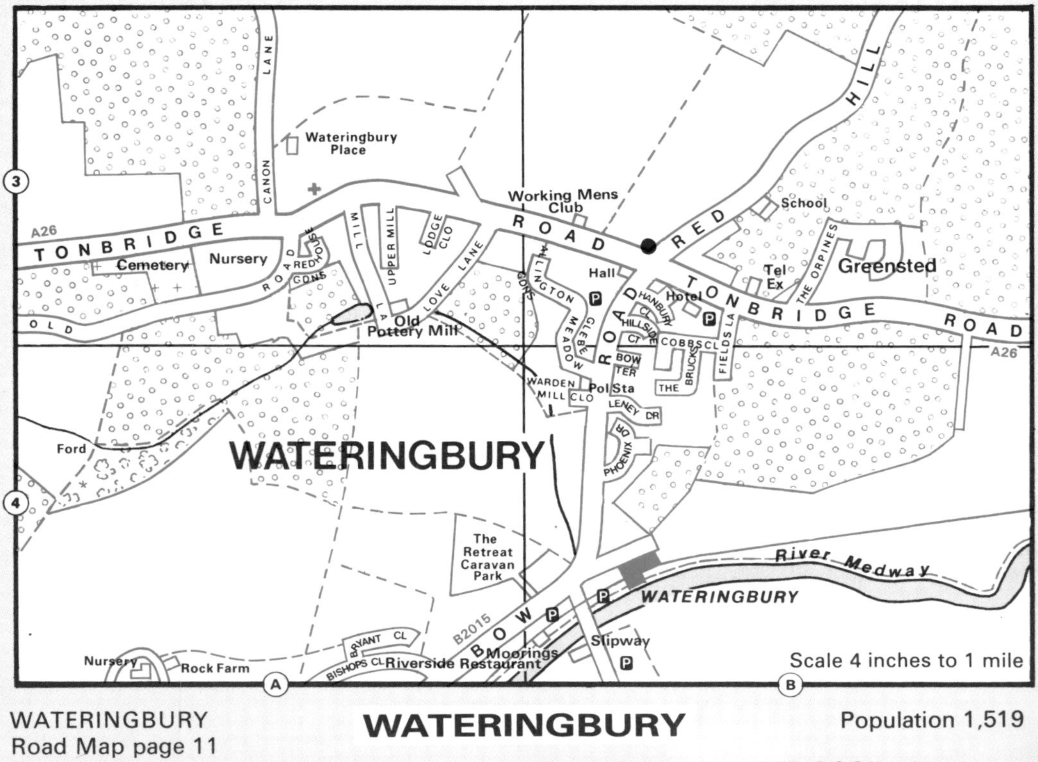

WATERINGBURY
Road Map page 11

WATERINGBURY

Population 1,519

MARGATE

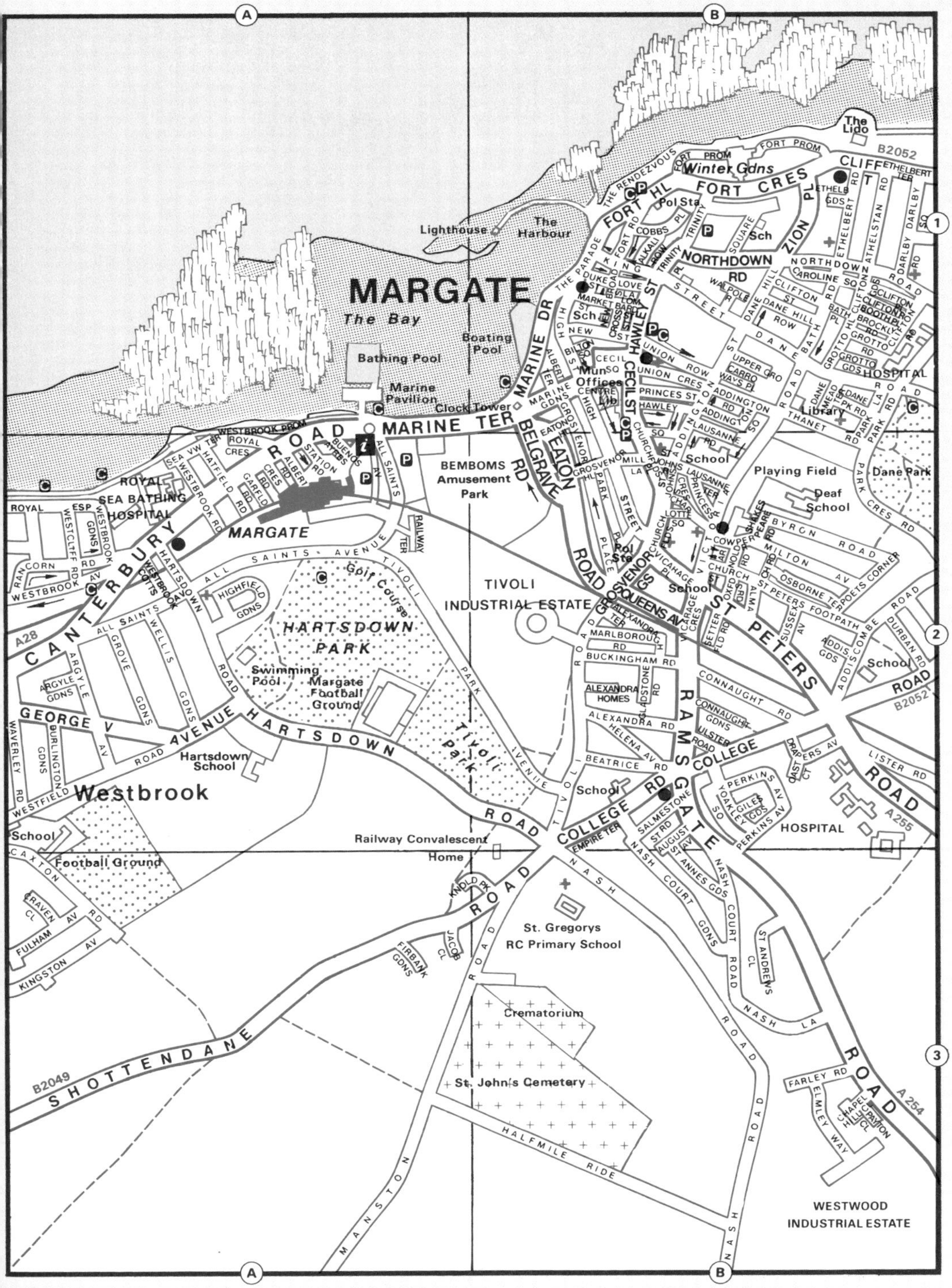

MEOPHAM

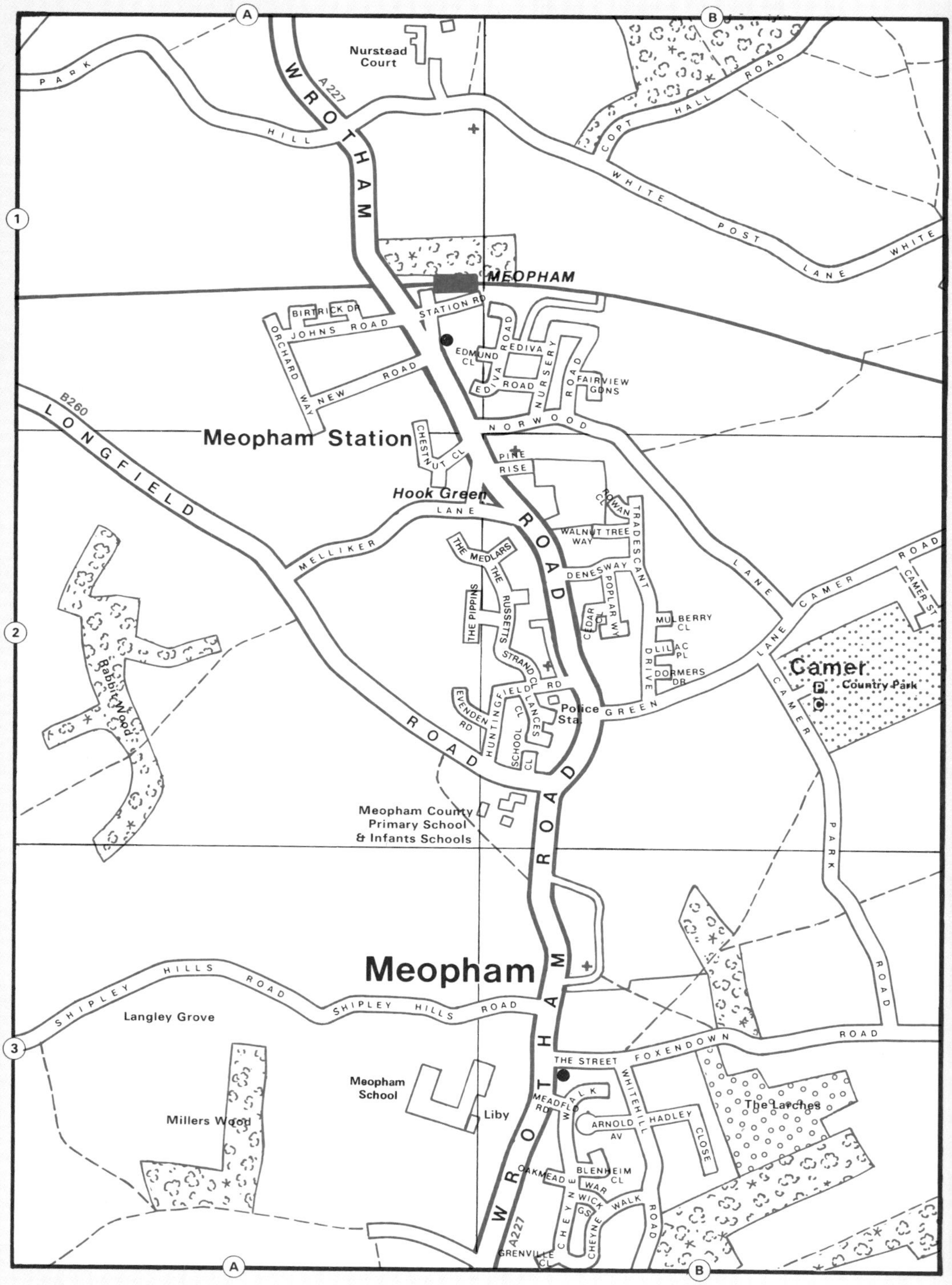

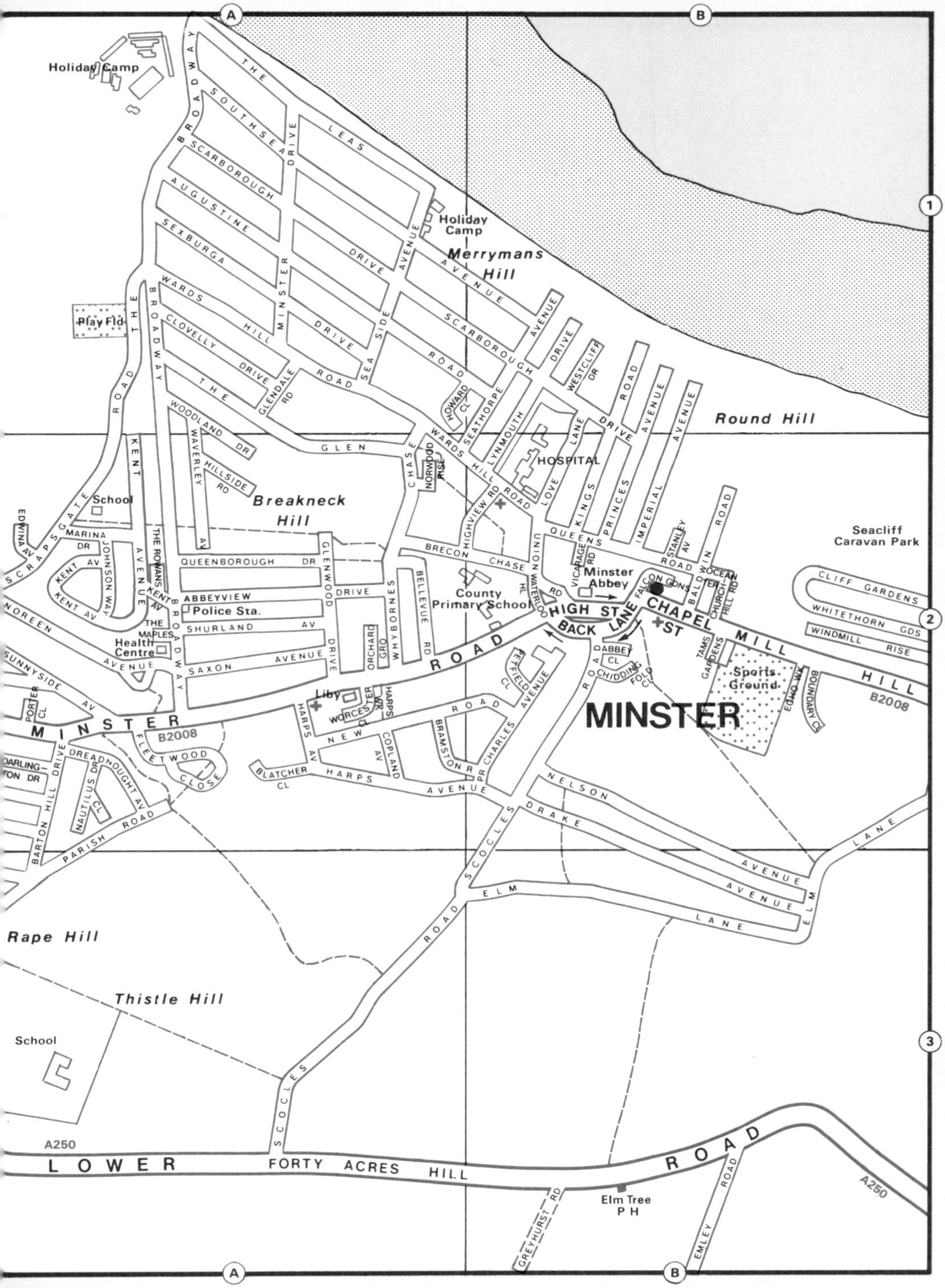

MINSTER-IN-SHEPPEY
Road Map page 12
For extended coverage of this area   —   see Estate Publications RED BOOK—Swale
Reproduction prohibited without prior permission   © Estate Publications

MINSTER-IN-THANET

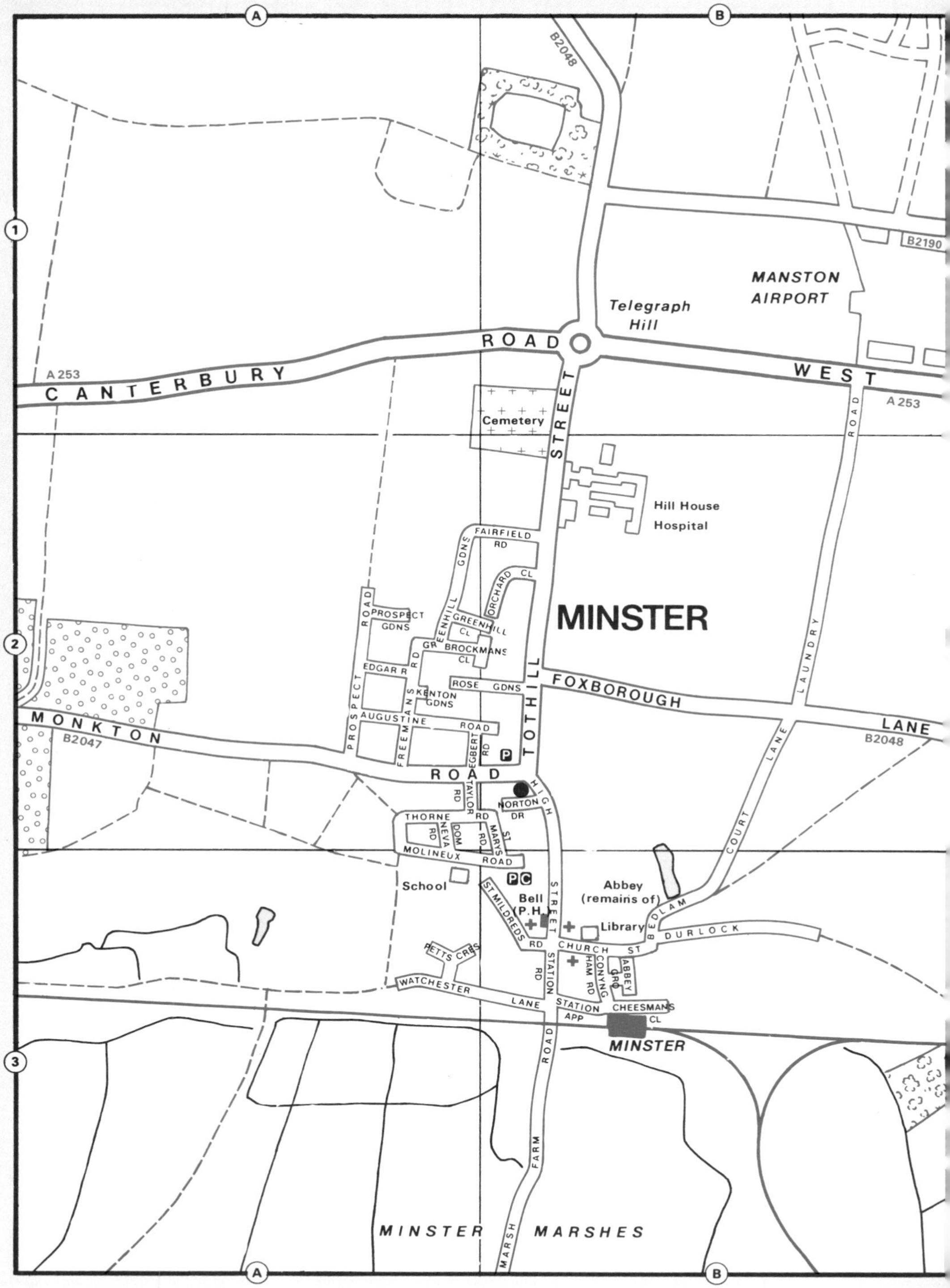

MINSTER-IN-THANET
Road Map page 13   Population 3,158
For extended coverage of this area—see Estate Publications RED BOOK—Thanet/Canterbury
Reproduction prohibited without prior permission   © Estate Publications

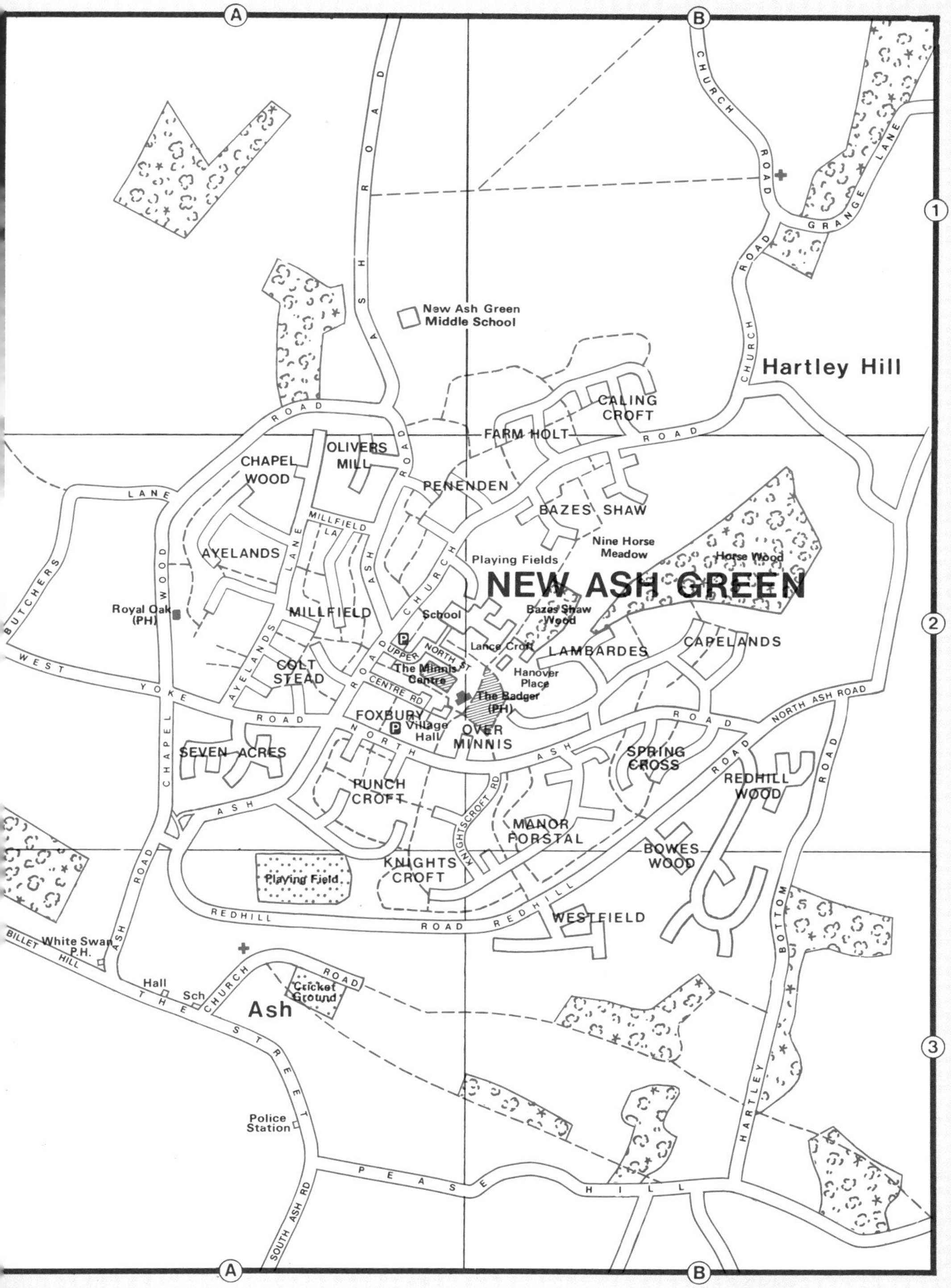

NEW ROMNEY

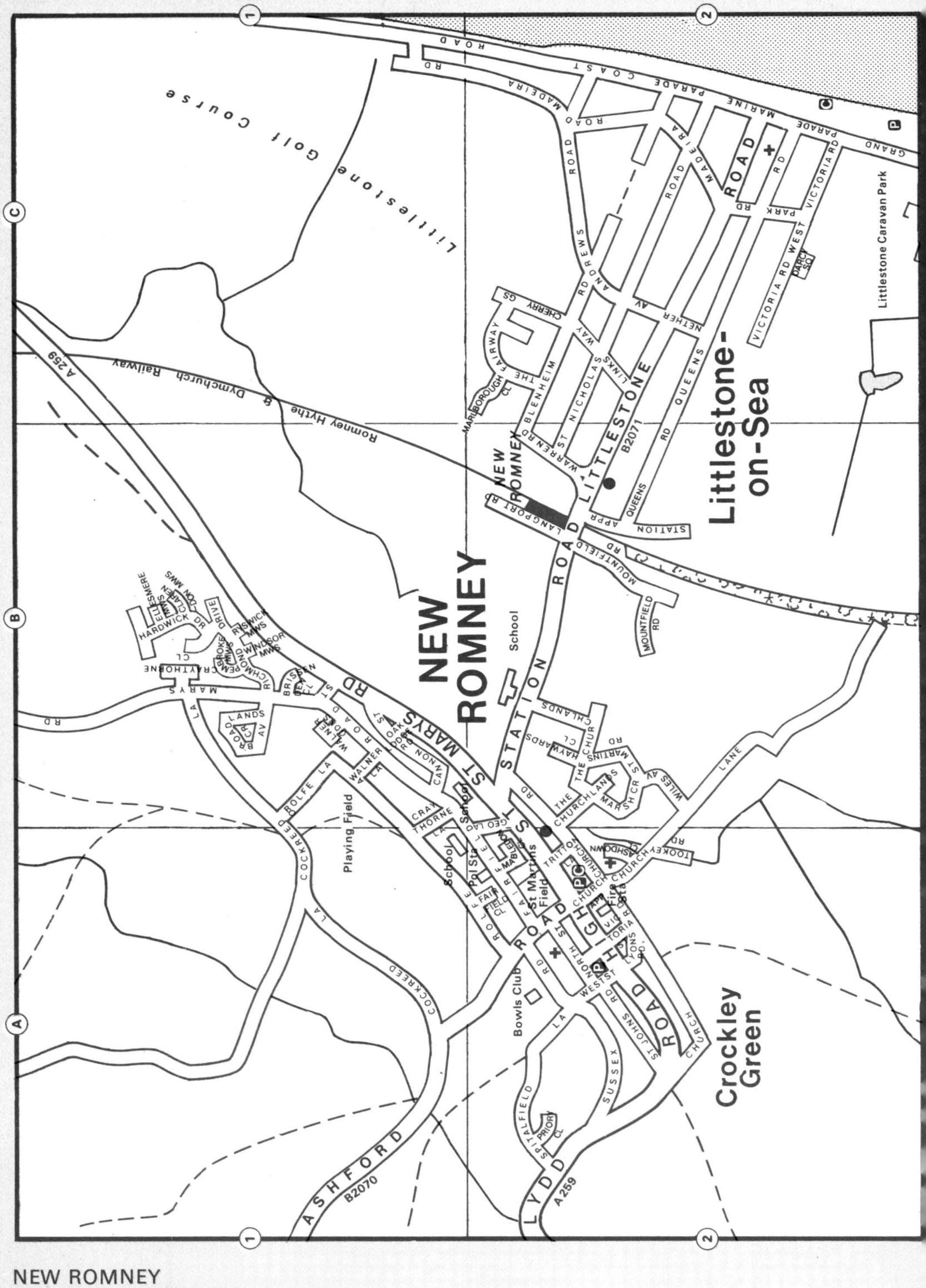

NEW ROMNEY
Road Map page 15   Population 4,547
For extended coverage of this area—see Estate Publications RED BOOK—Rye & Romney Marsh
Reproduction prohibited without prior permission   © Estate Publications

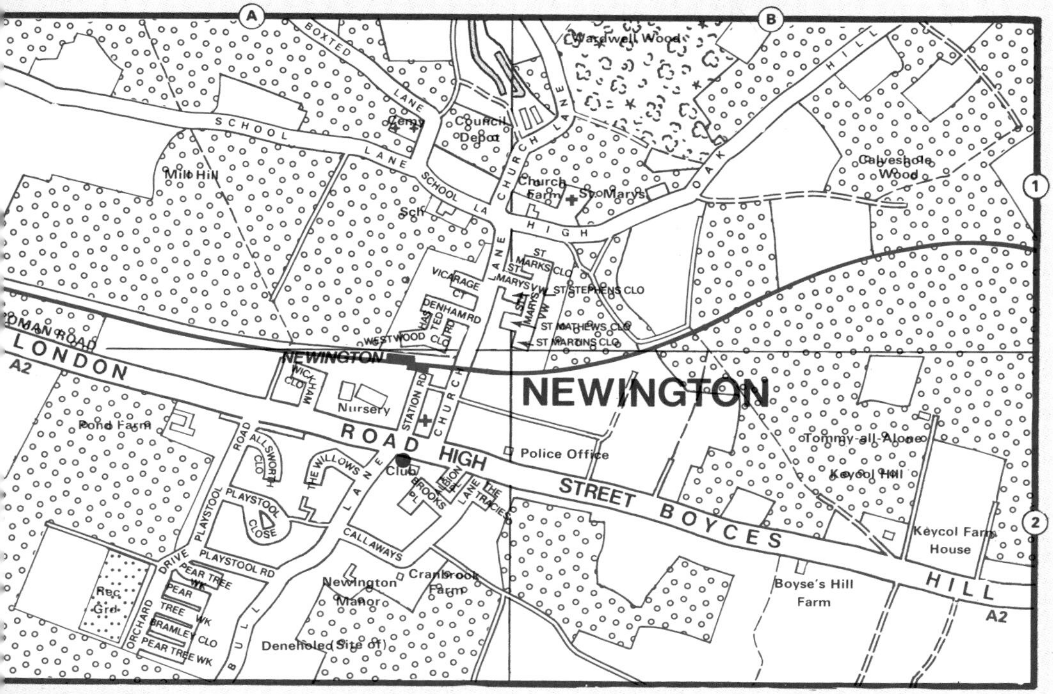

EWINGTON
oad Map page 13   Population 2,538

Scale 4 inches to 1 mile

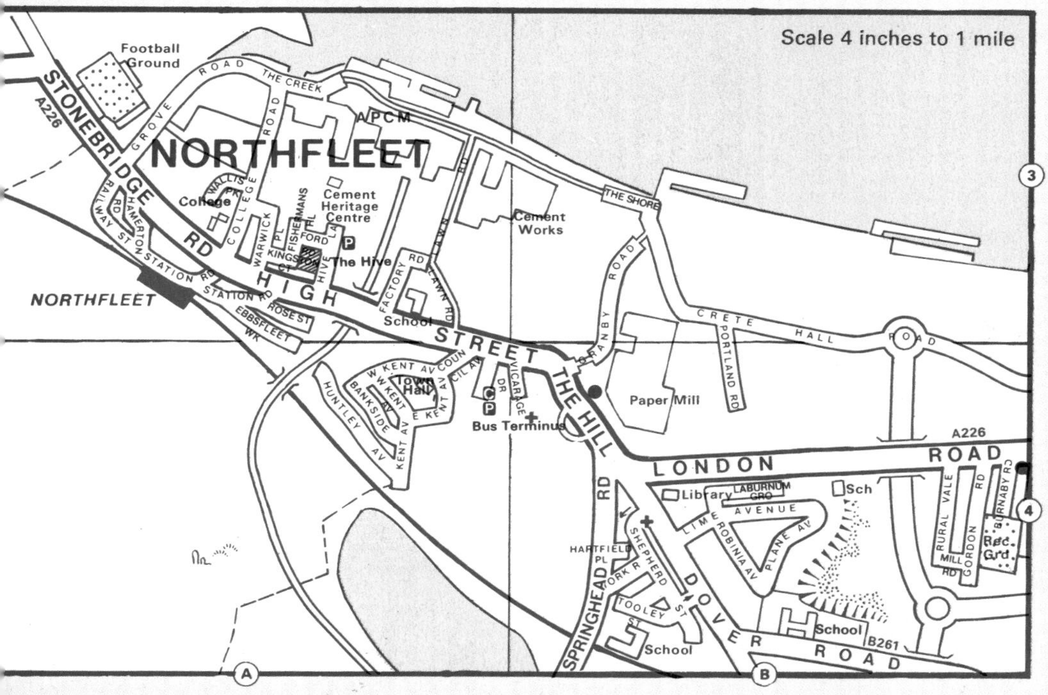

NORTHFLEET

Population 26,310

or extended coverage of this area—see Estate Publications RED BOOK—Gravesend/Dartford

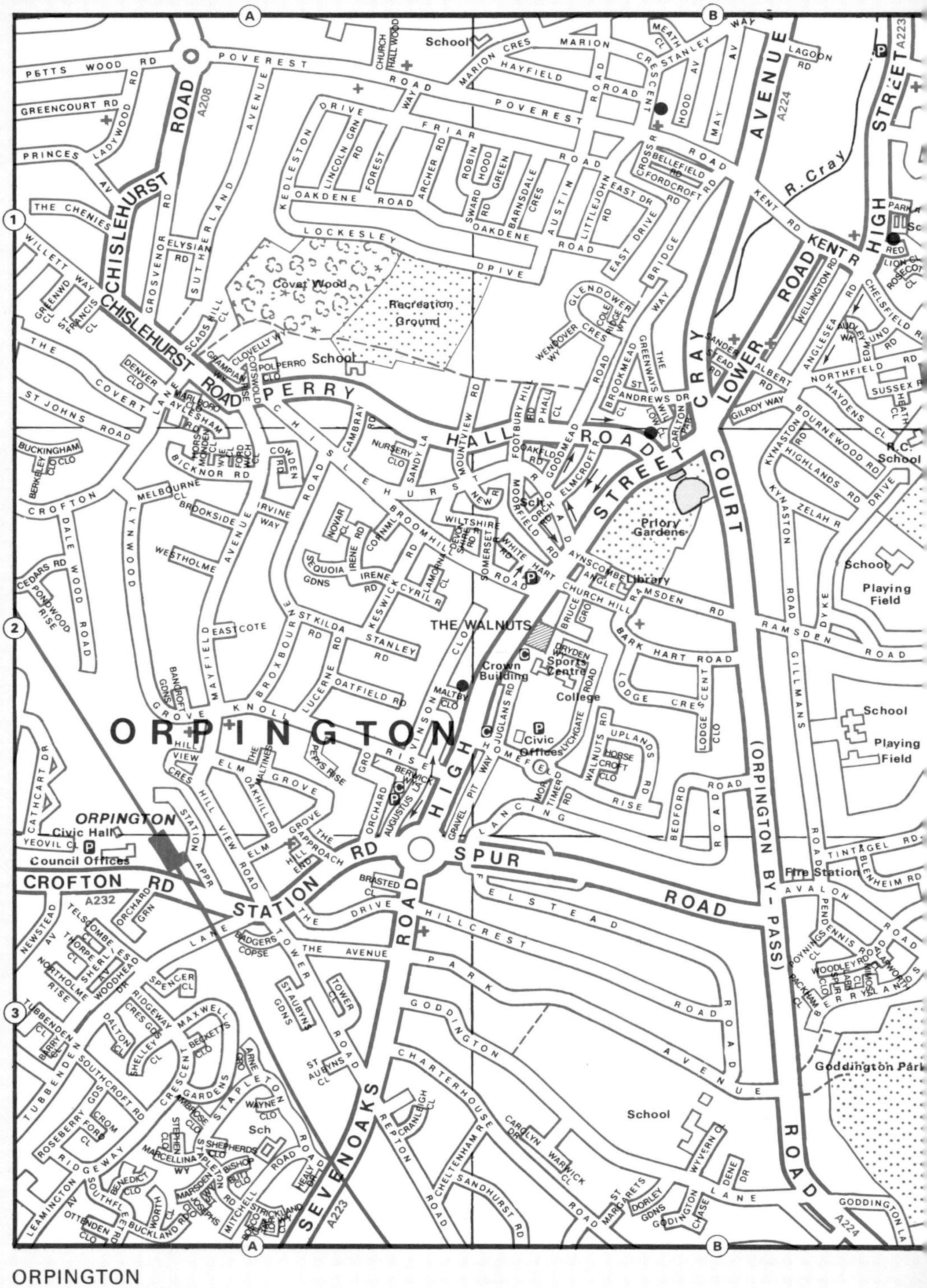

ORPINGTON
Road Map page 10
For extended coverage of this area   —   see Estate Publications RED BOOK—Bromley
Reproduction prohibited without prior permission   © Estate Publications

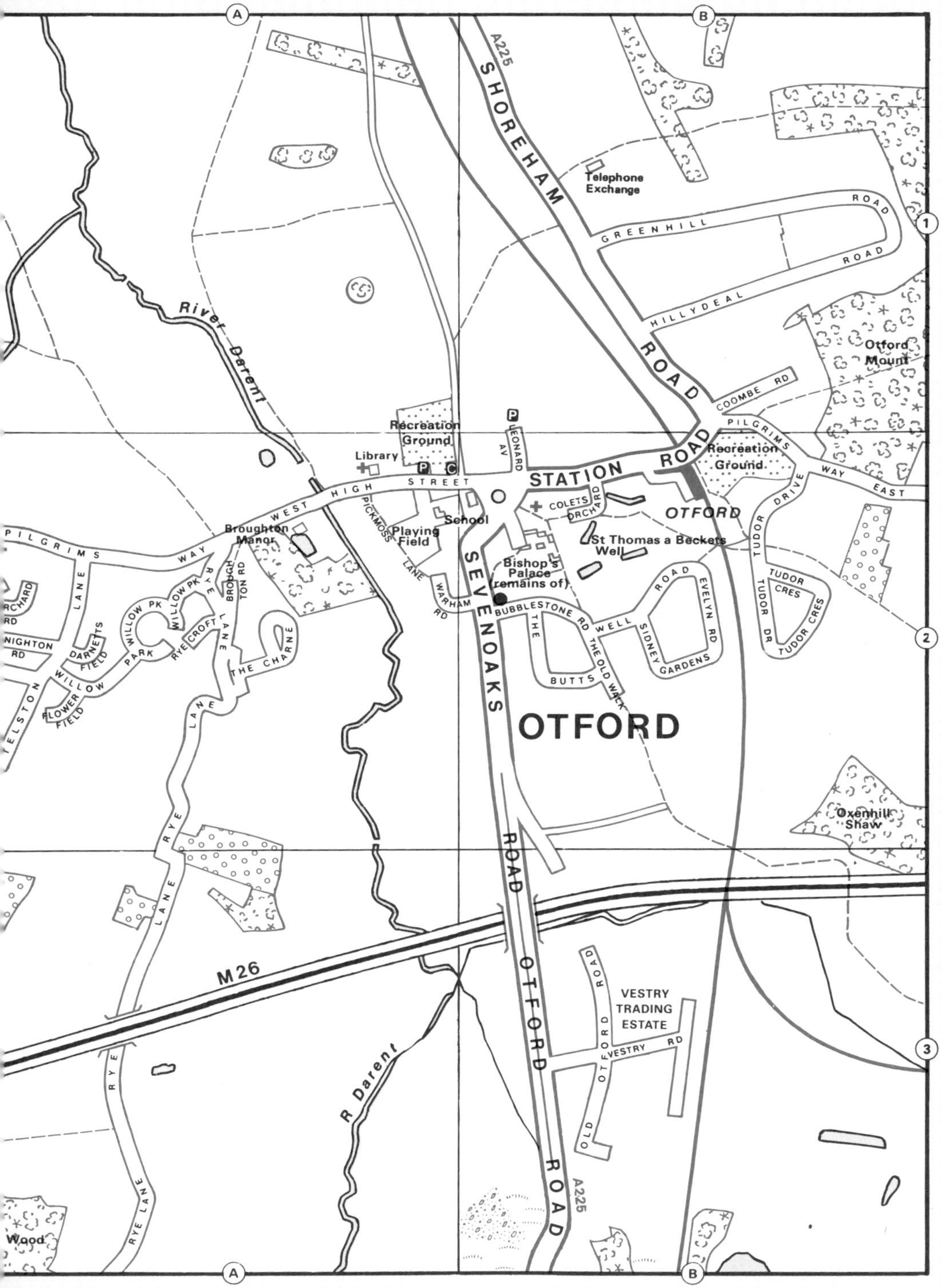

PADDOCK WOOD

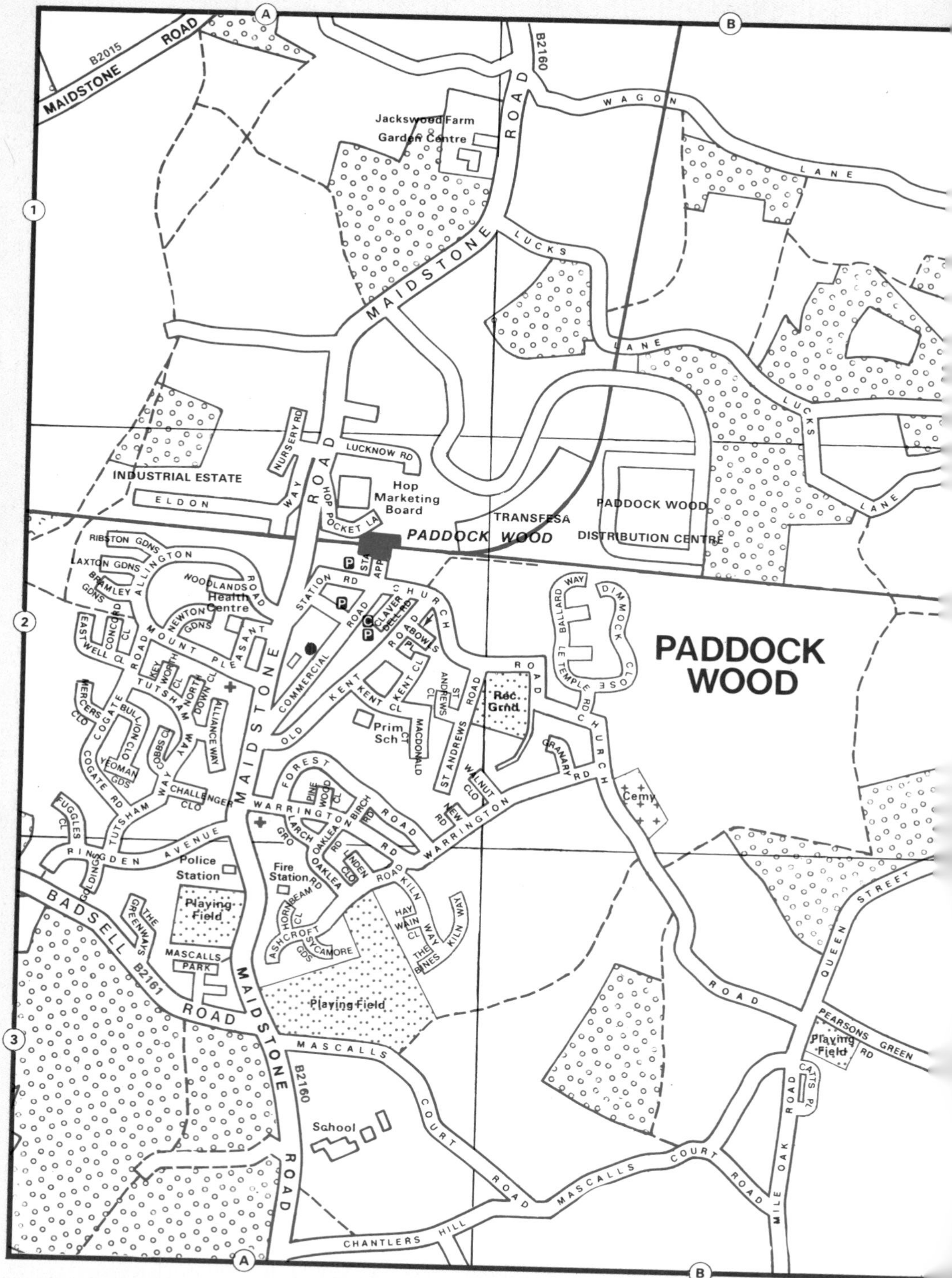

PADDOCK WOOD
Road Map page 14   Population 6,490
For extended coverage of this area— see Estate Publications RED BOOK—Tunbridge Wells/
Reproduction prohibited without prior permission   Tonbridge

© Estate Publications

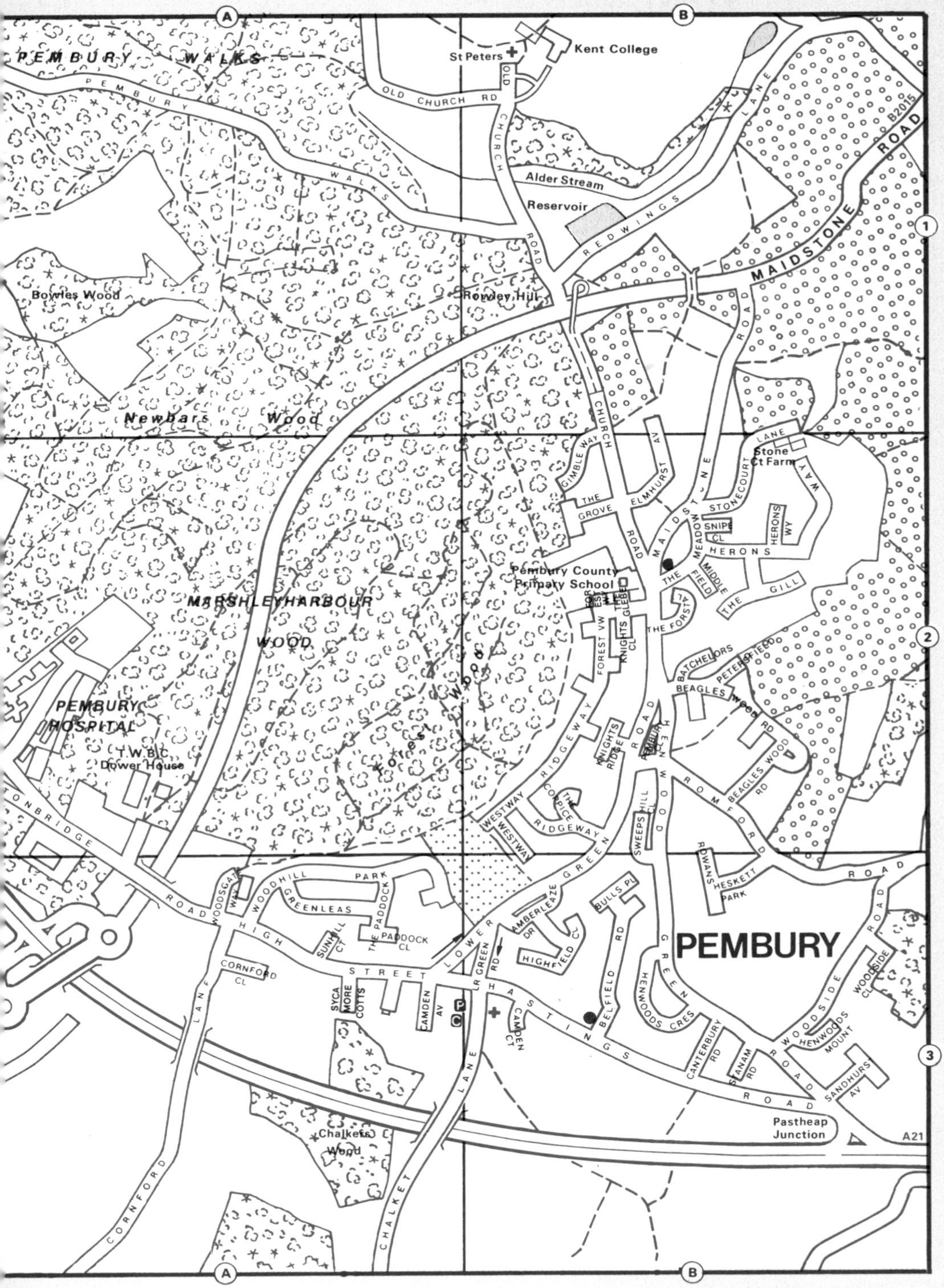

PEMBURY
Road Map page 14   Population 6,285
For extended coverage of this area— see Estate Publications RED BOOK— Tunbridge Wells/
Reproduction prohibited without prior permission   Tonbridge

© Estate Publications

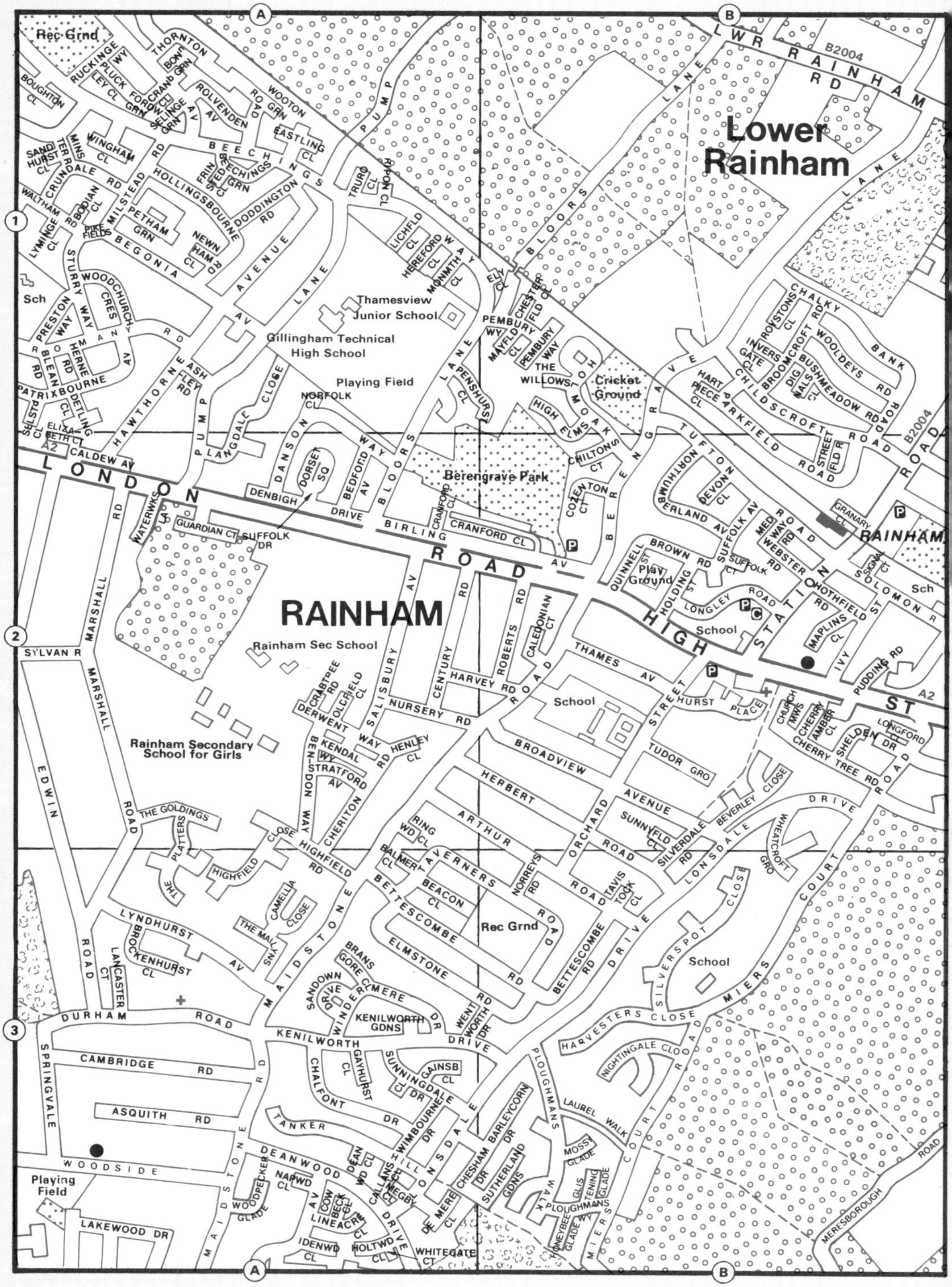

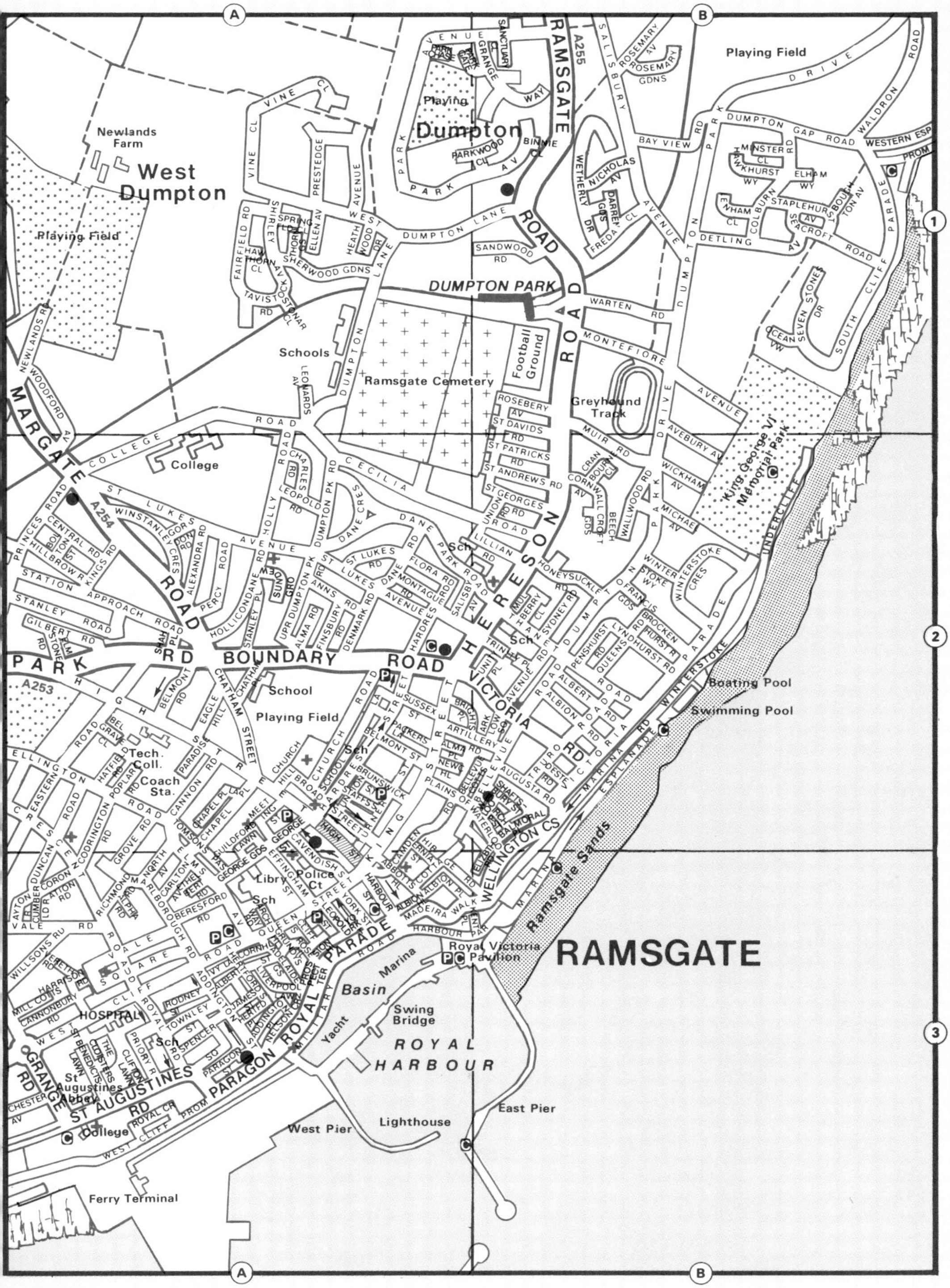

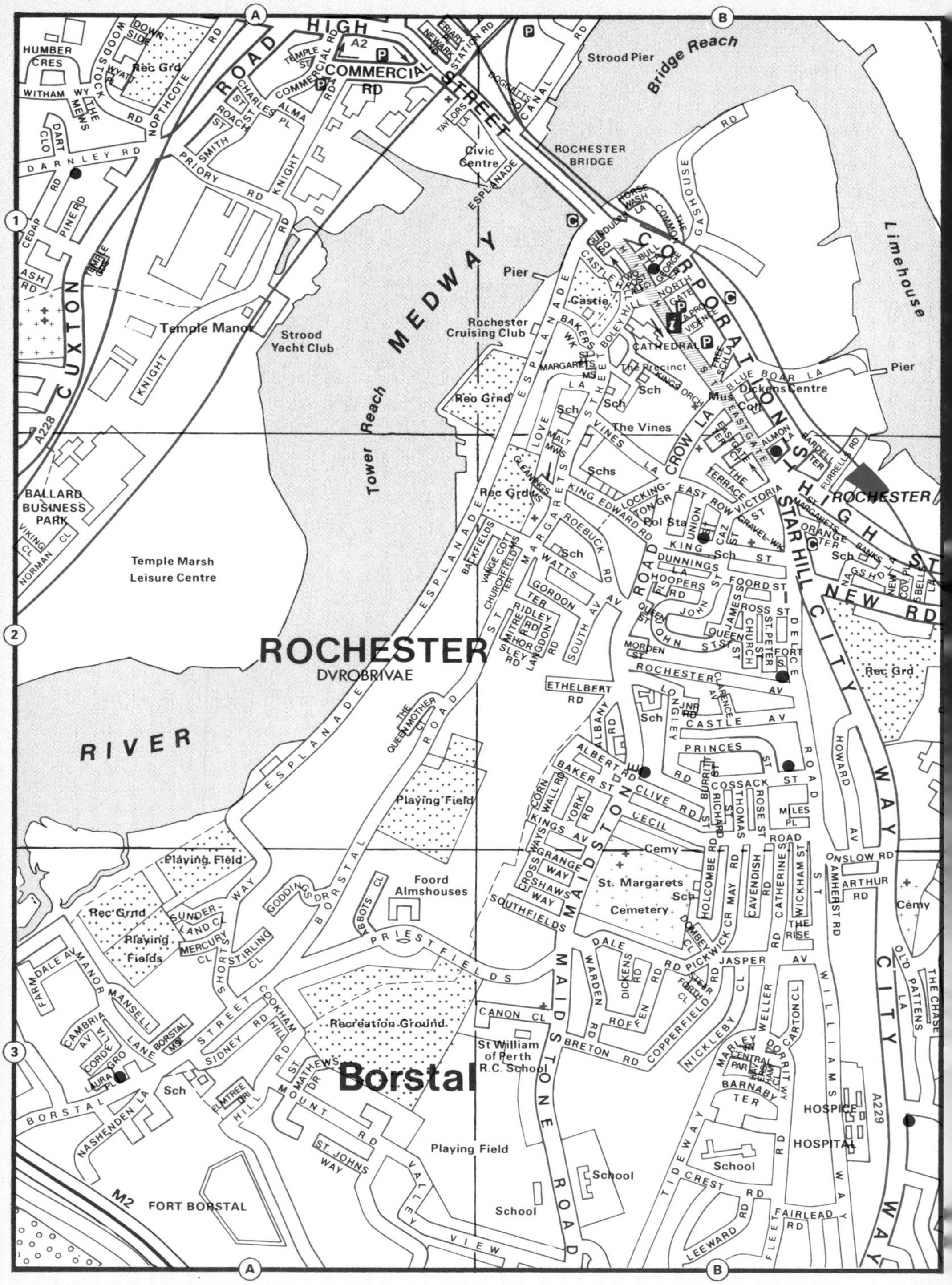

ROCHESTER
Road Map page 11   Population 67,651
For extended coverage of this area—see Estate Publications RED BOOK—Medway/Gillingham
Reproduction prohibited without prior permission   © Estate Publications

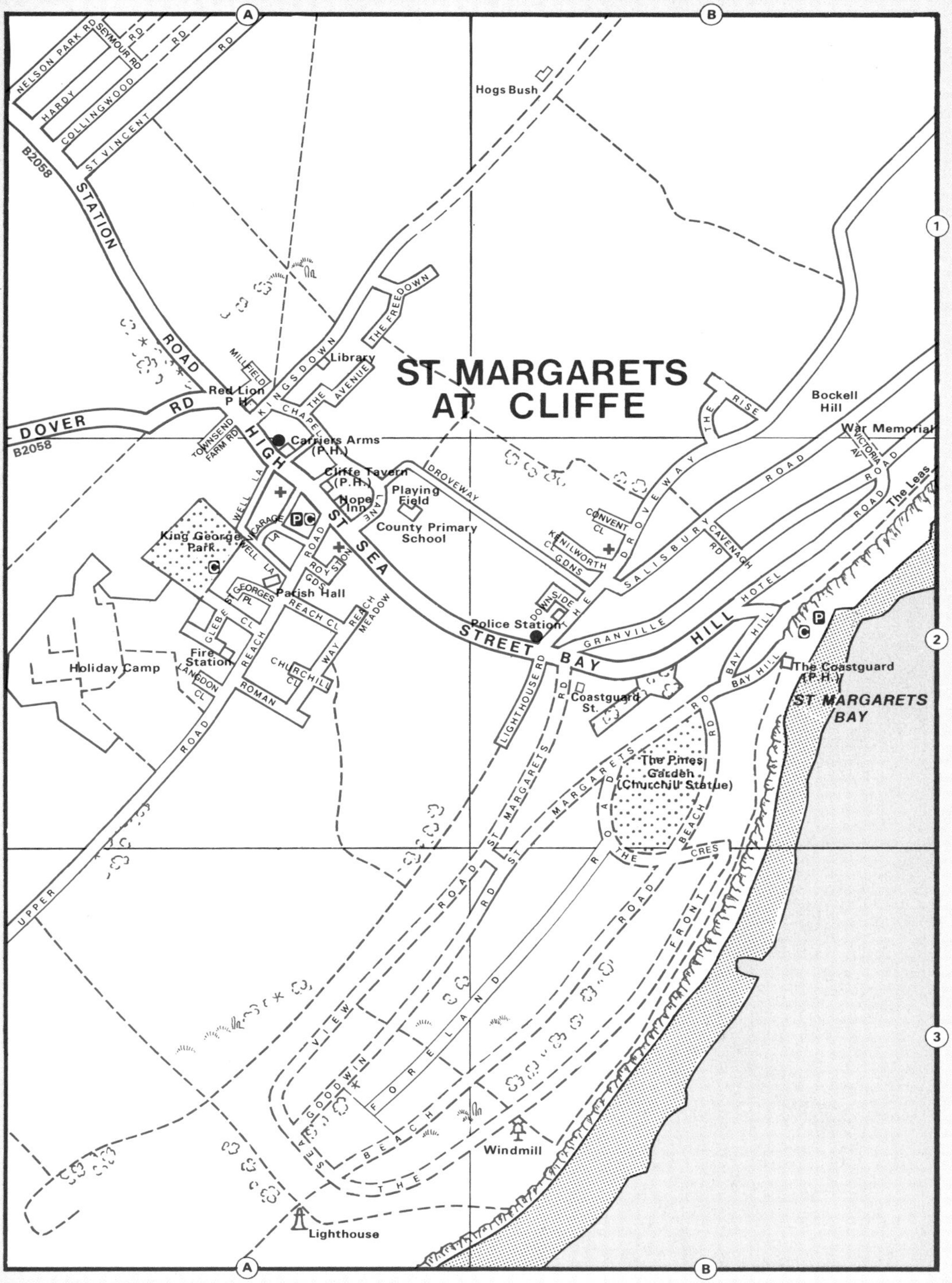

ST. MARGARETS AT CLIFFE

ST. MARGARET'S-AT-CLIFFE
Road Map page 13   Population 2,312
For extended coverage of this area—see Estate Publications RED BOOK—Folkestone/Dover

ST. MARY'S BAY

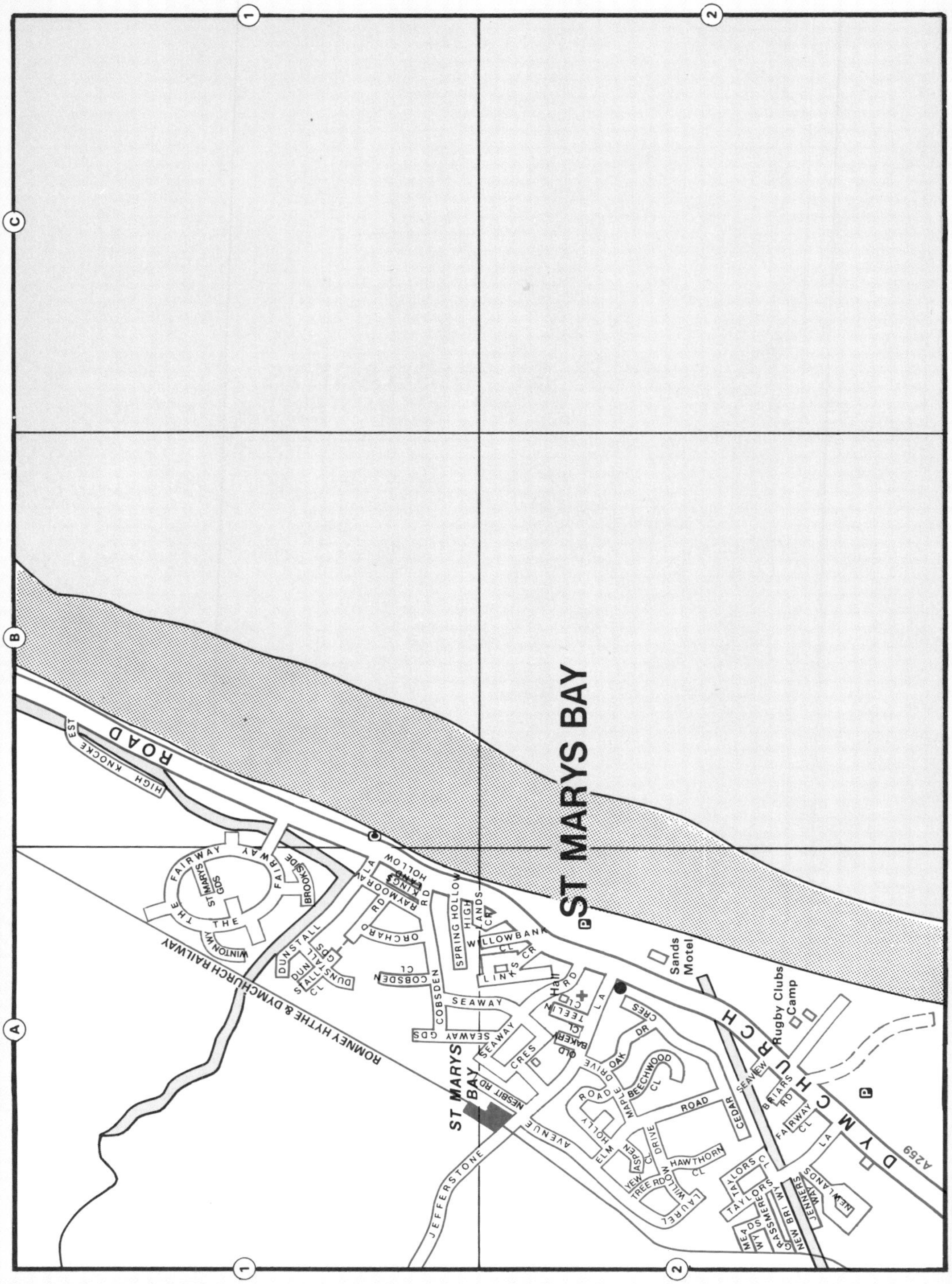

ST. MARY'S BAY

For extended coverage of this area—see Estate Publications RED BOOK—Rye & Romney Marsh
© Estate Publications

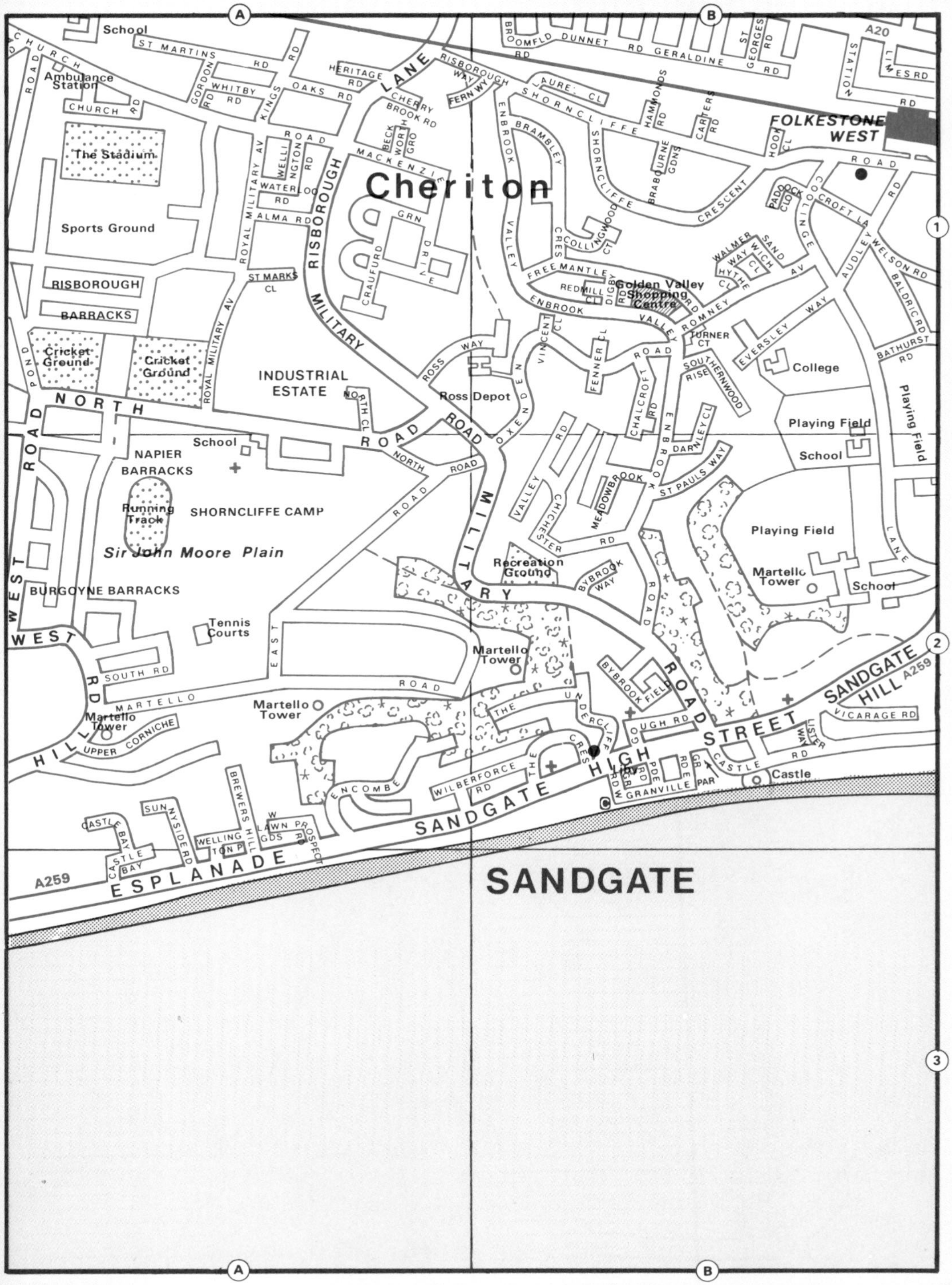

SANDGATE

For extended coverage of this area—see Estate Publications RED BOOK—Folkestone/Dover

SANDWICH

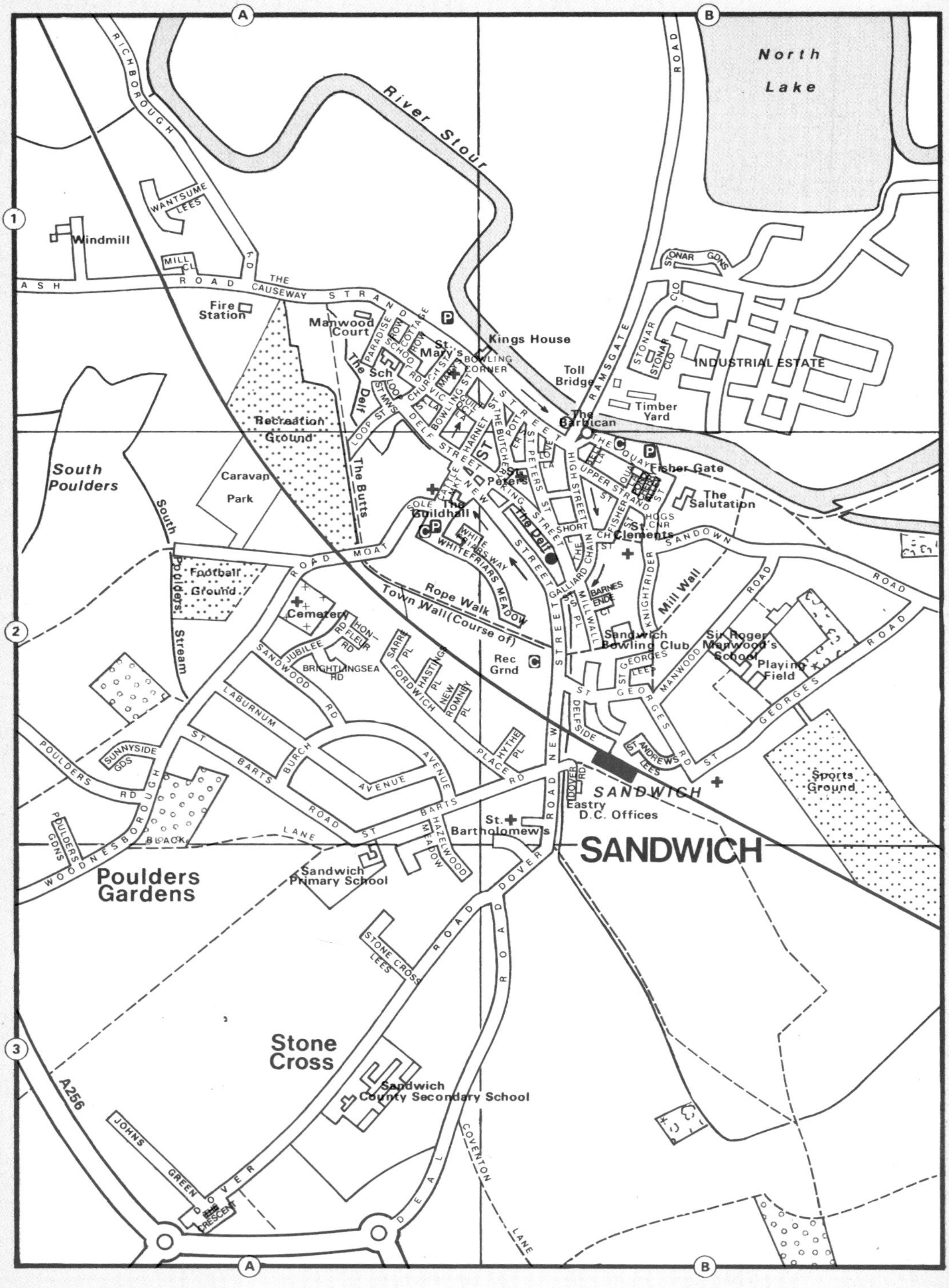

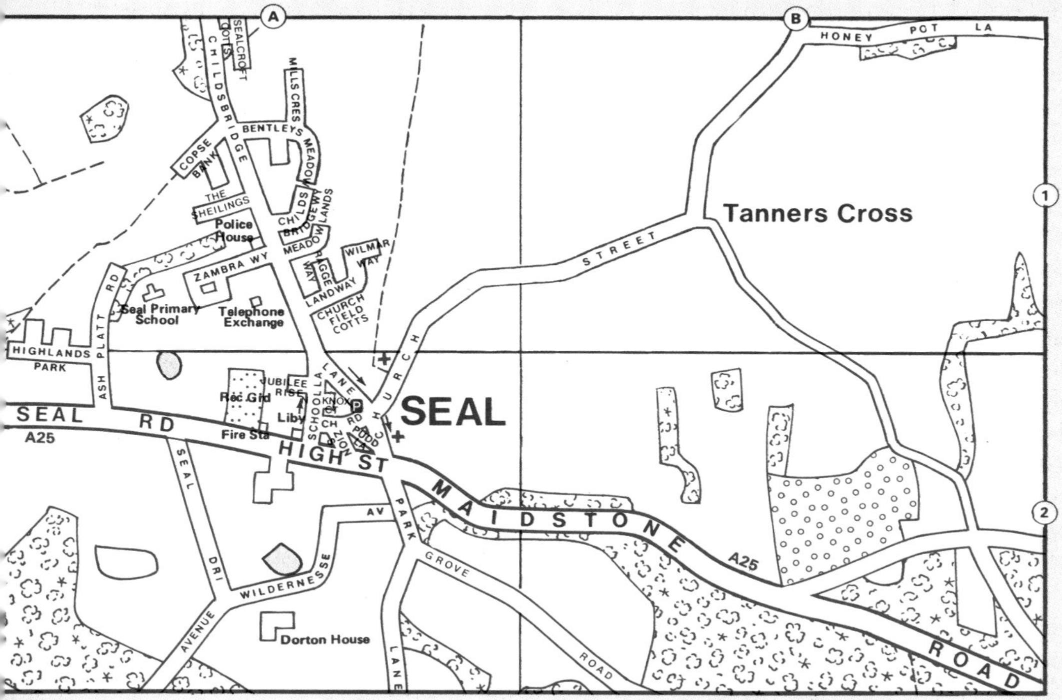

Tanners Cross

SEAL

SEAL
Road Map page 10   Population 2,423

For extended coverage of this area —
see Estate Publications RED BOOK—Sevenoaks

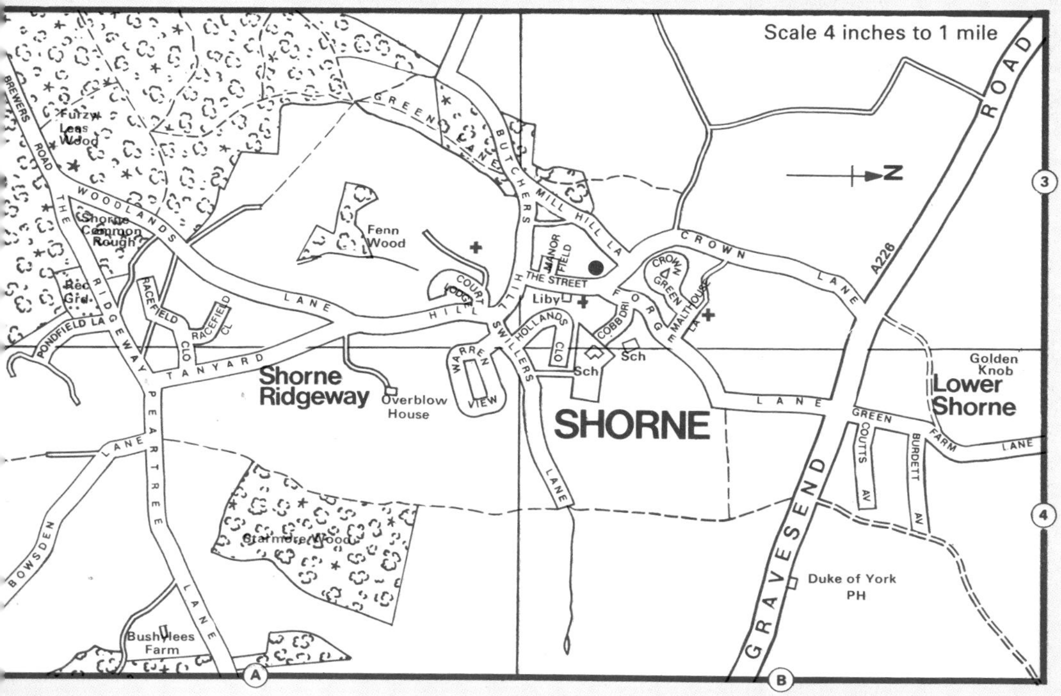

Scale 4 inches to 1 mile

Shorne Ridgeway

SHORNE

Lower Shorne

SHORNE
Road Map page 11
Reproduction prohibited without prior permission

SHORNE

Population 2,565

© Estate Publications

SEVENOAKS

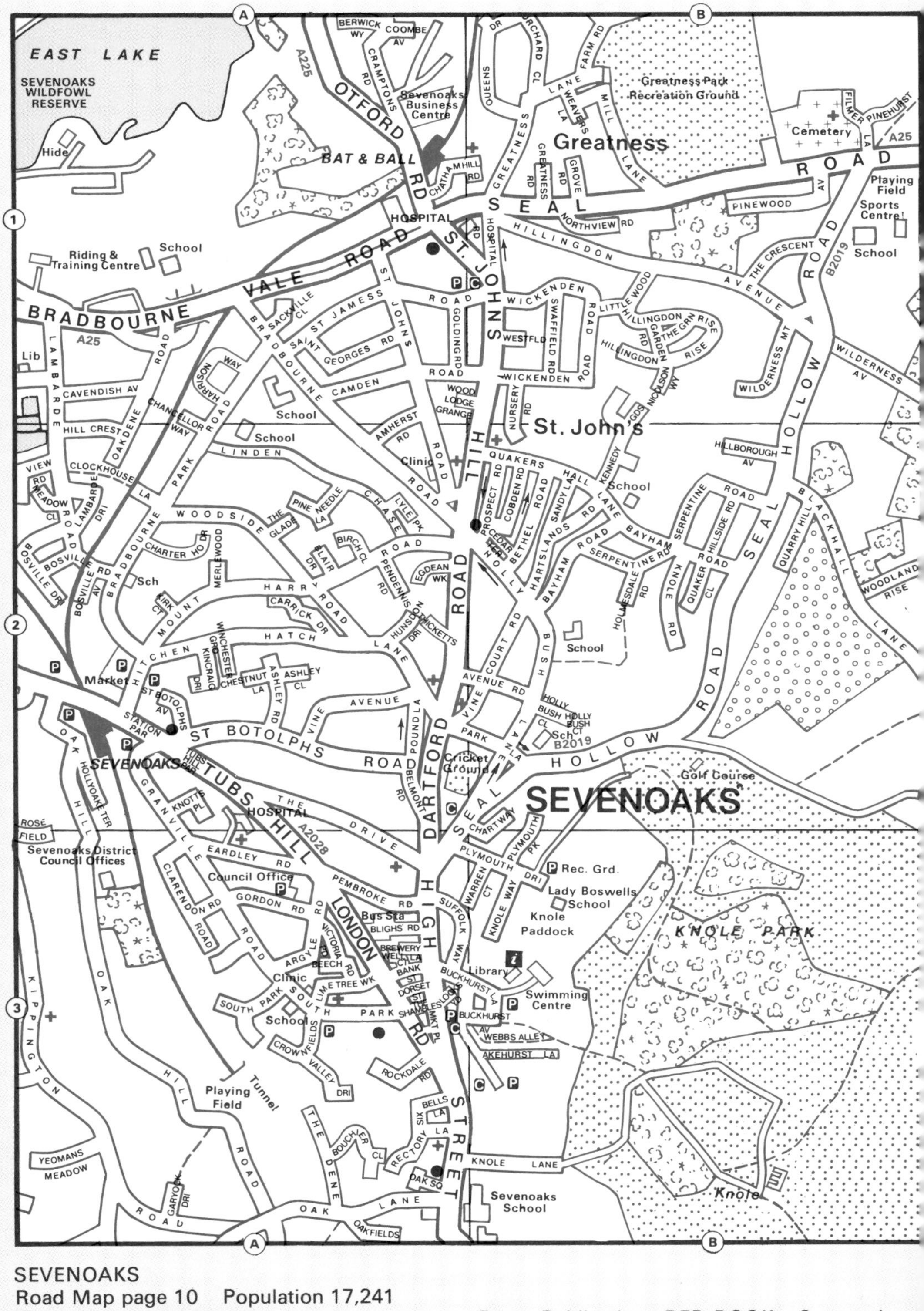

SEVENOAKS
Road Map page 10   Population 17,241
For extended coverage of this area   —   see Estate Publications RED BOOK—Sevenoaks
Reproduction prohibited without prior permission
© Estate Publications

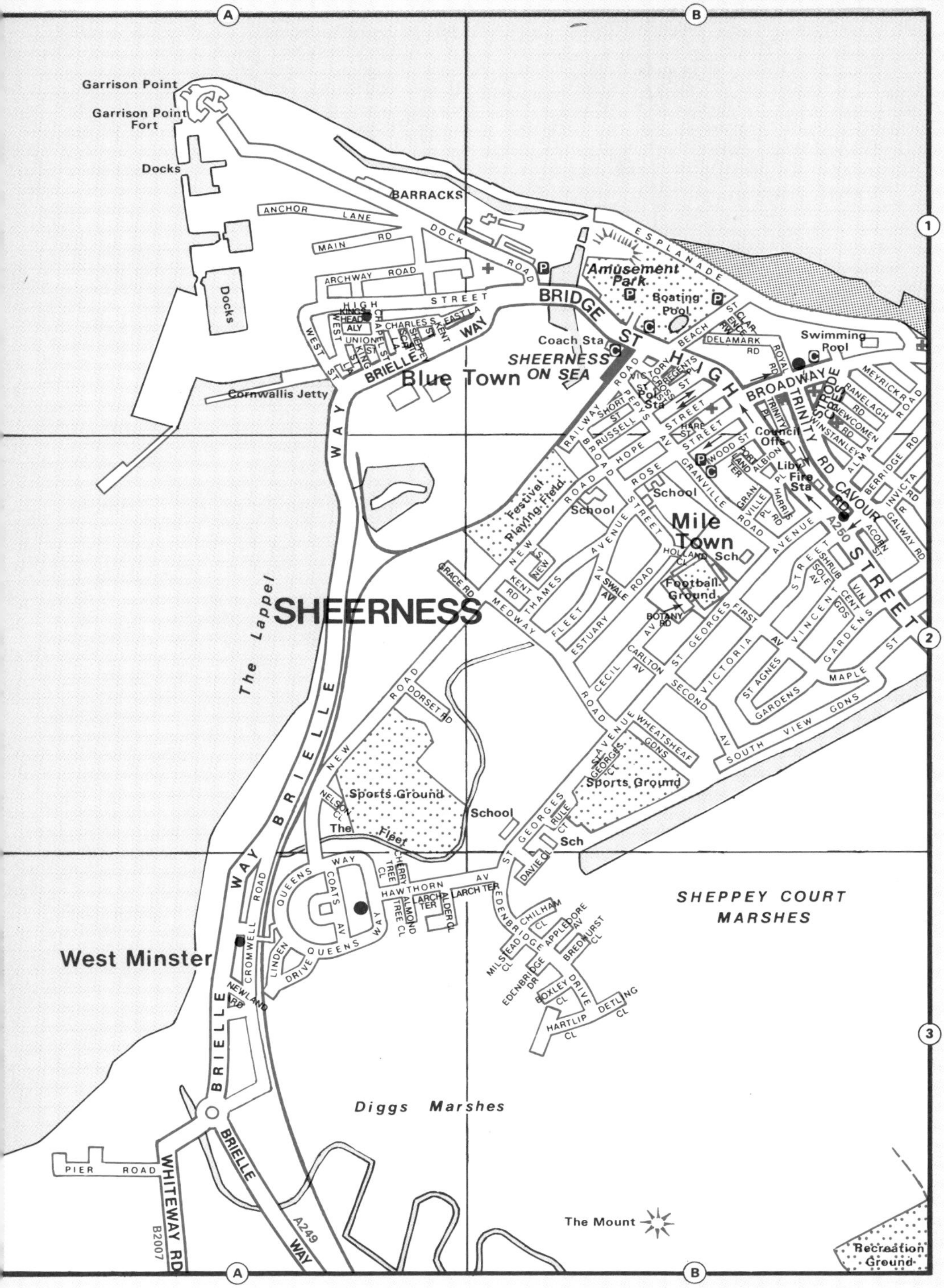

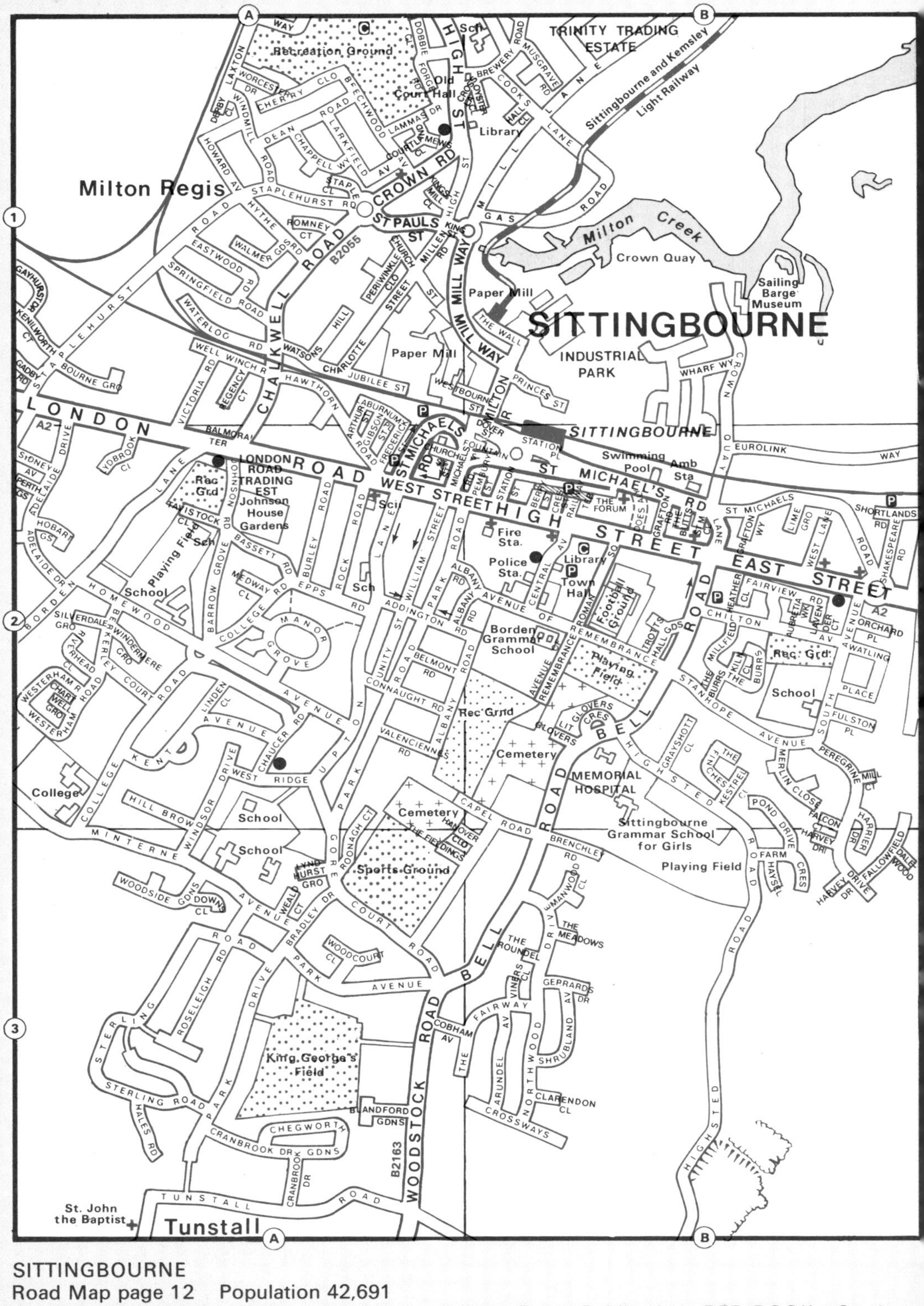

SITTINGBOURNE
Road Map page 12   Population 42,691
For extended coverage of this area   —   see Estate Publications RED BOOK—Swale
Reproduction prohibited without prior permission   © Estate Publications

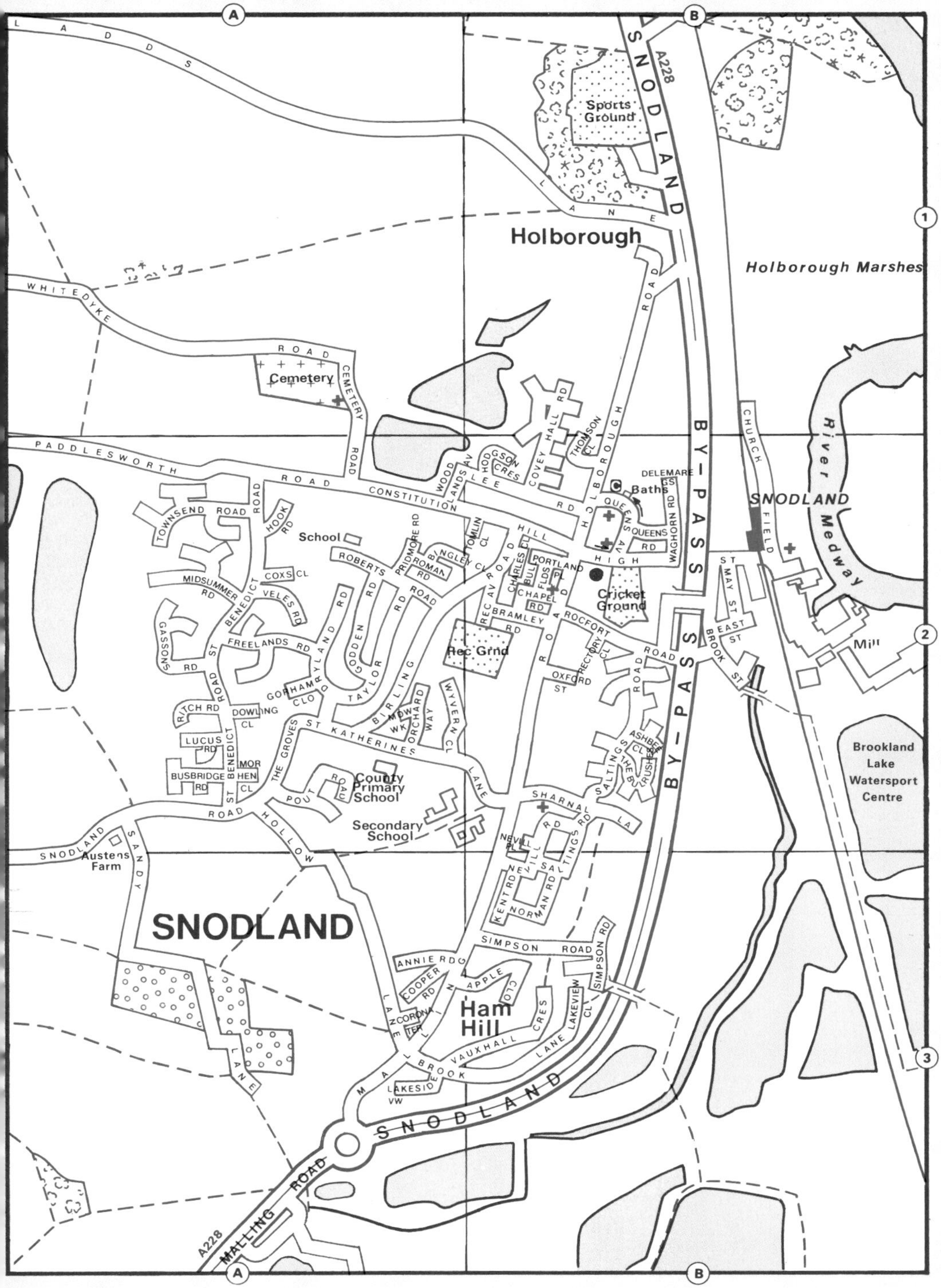

SNODLAND
Road Map page 11   Population 5,836
For extended coverage of this area — see Estate Publications RED BOOK—Maidstone

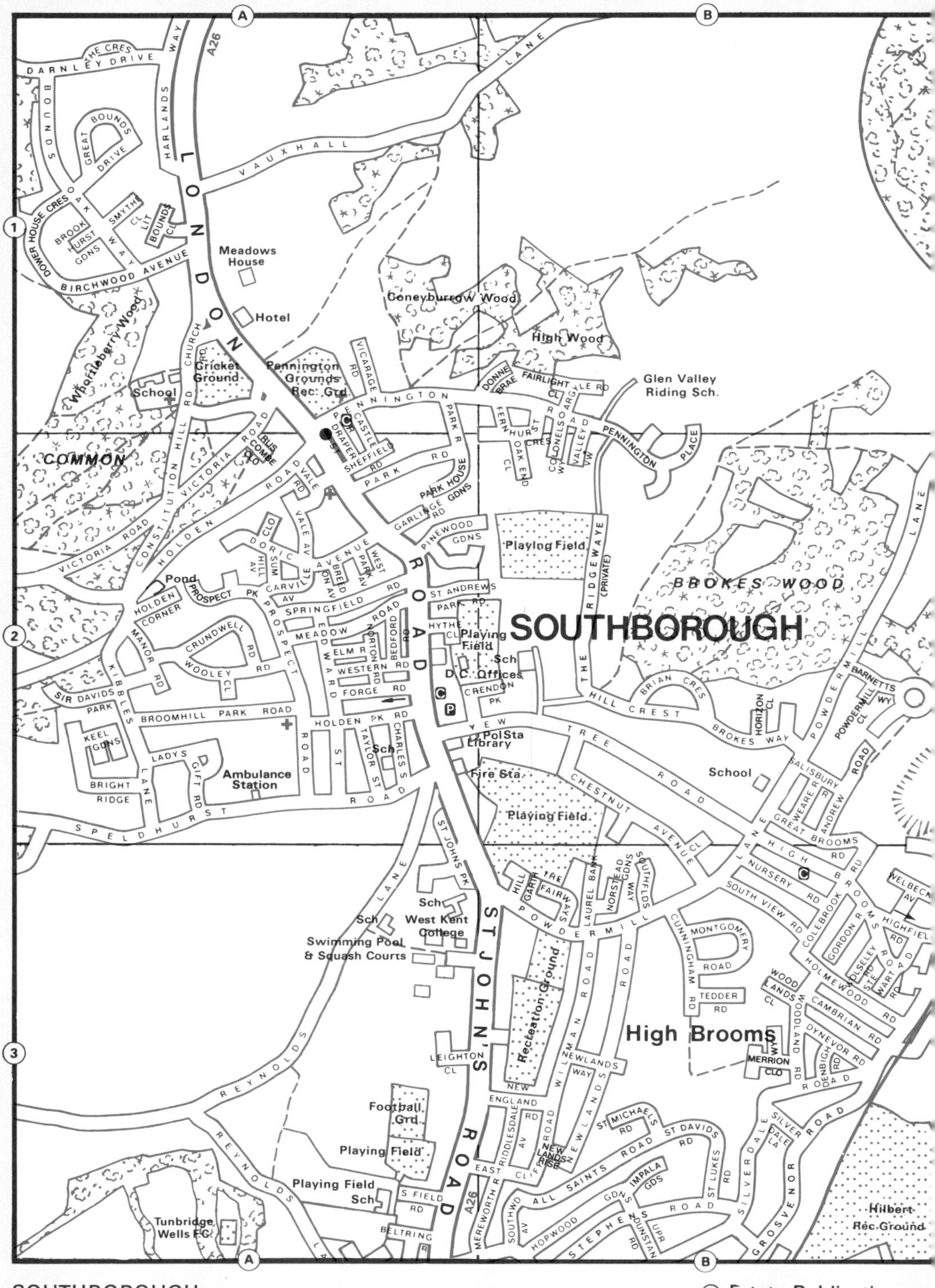

For extended coverage of this area— see Estate Publications RED BOOK— Tunbridge Wells/
Tonbridge

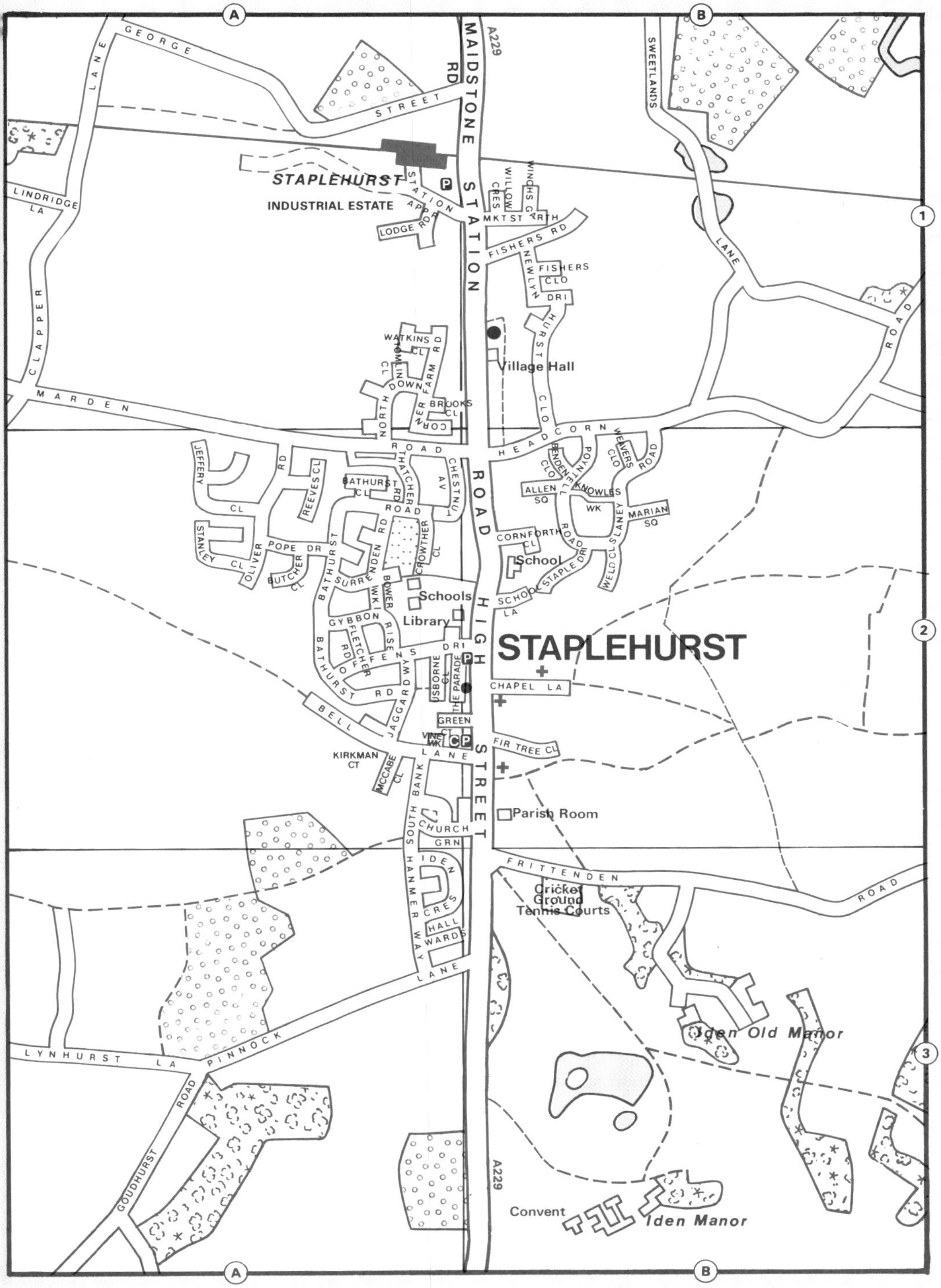

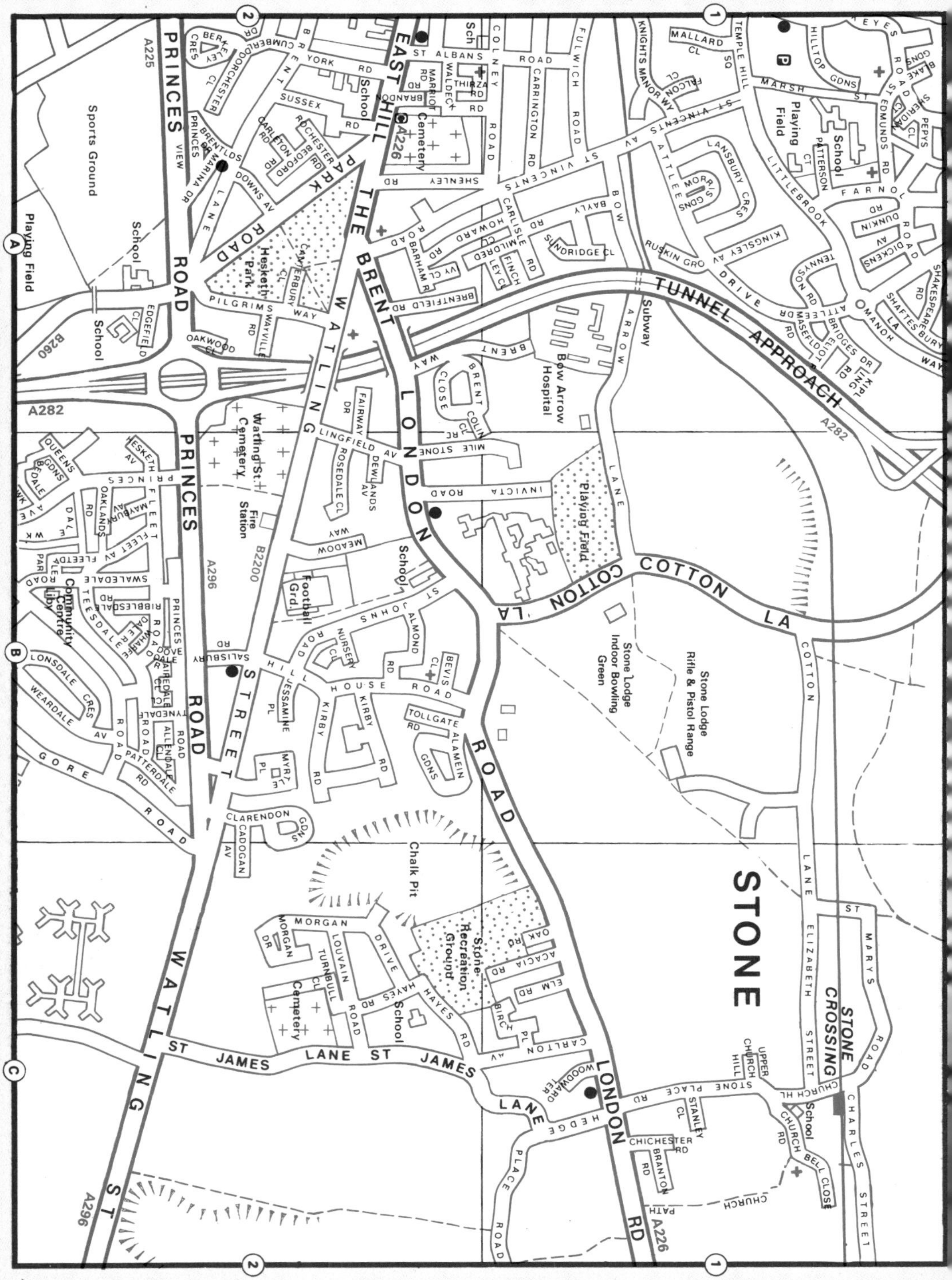

STONE
Road Map page 10   Population 8,560
For extended coverage of this area—see Estate Publications RED BOOK—Gravesend/Dartford
Reproduction prohibited without prior permission
© Estate Publications

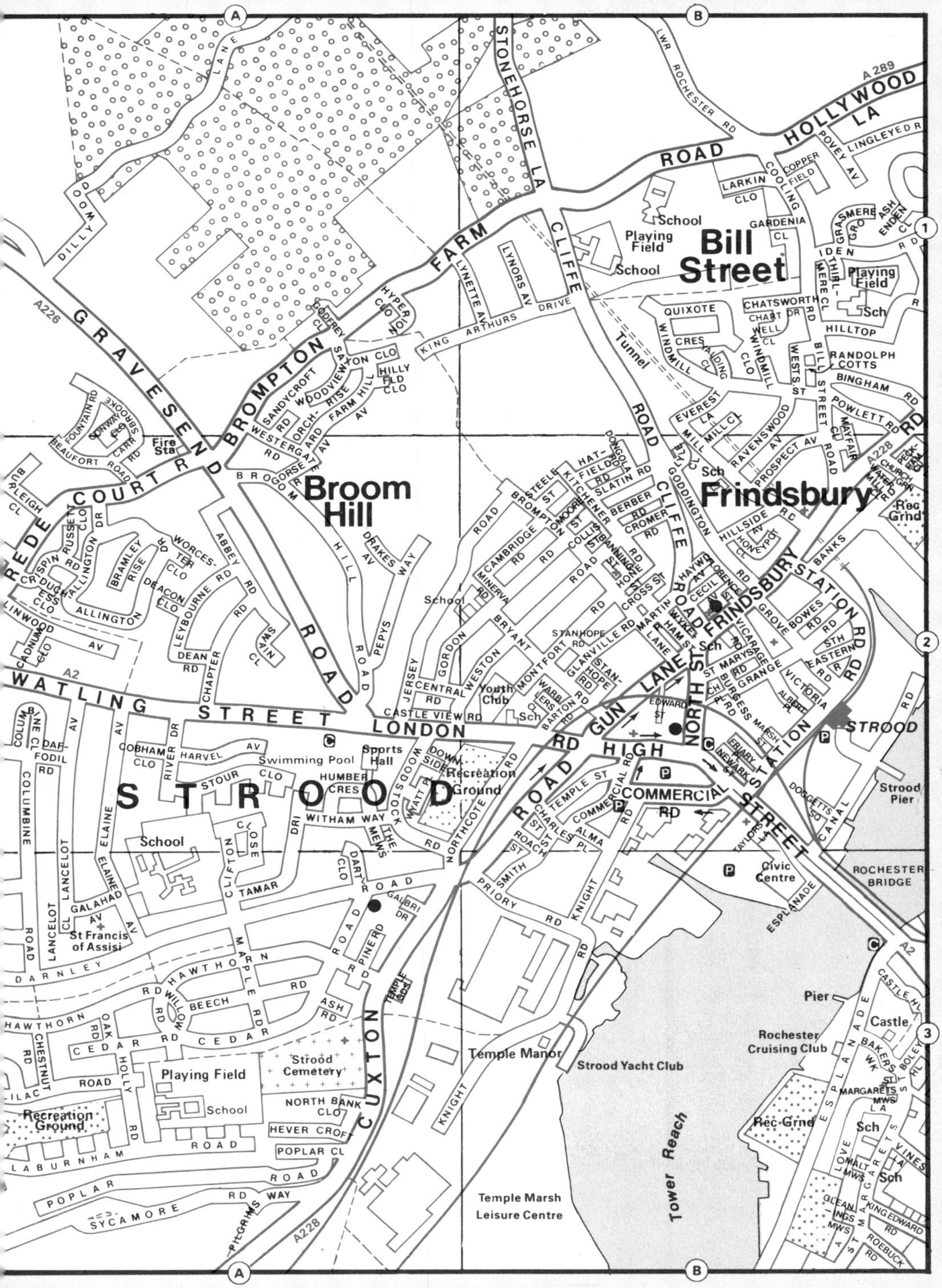

STURRY

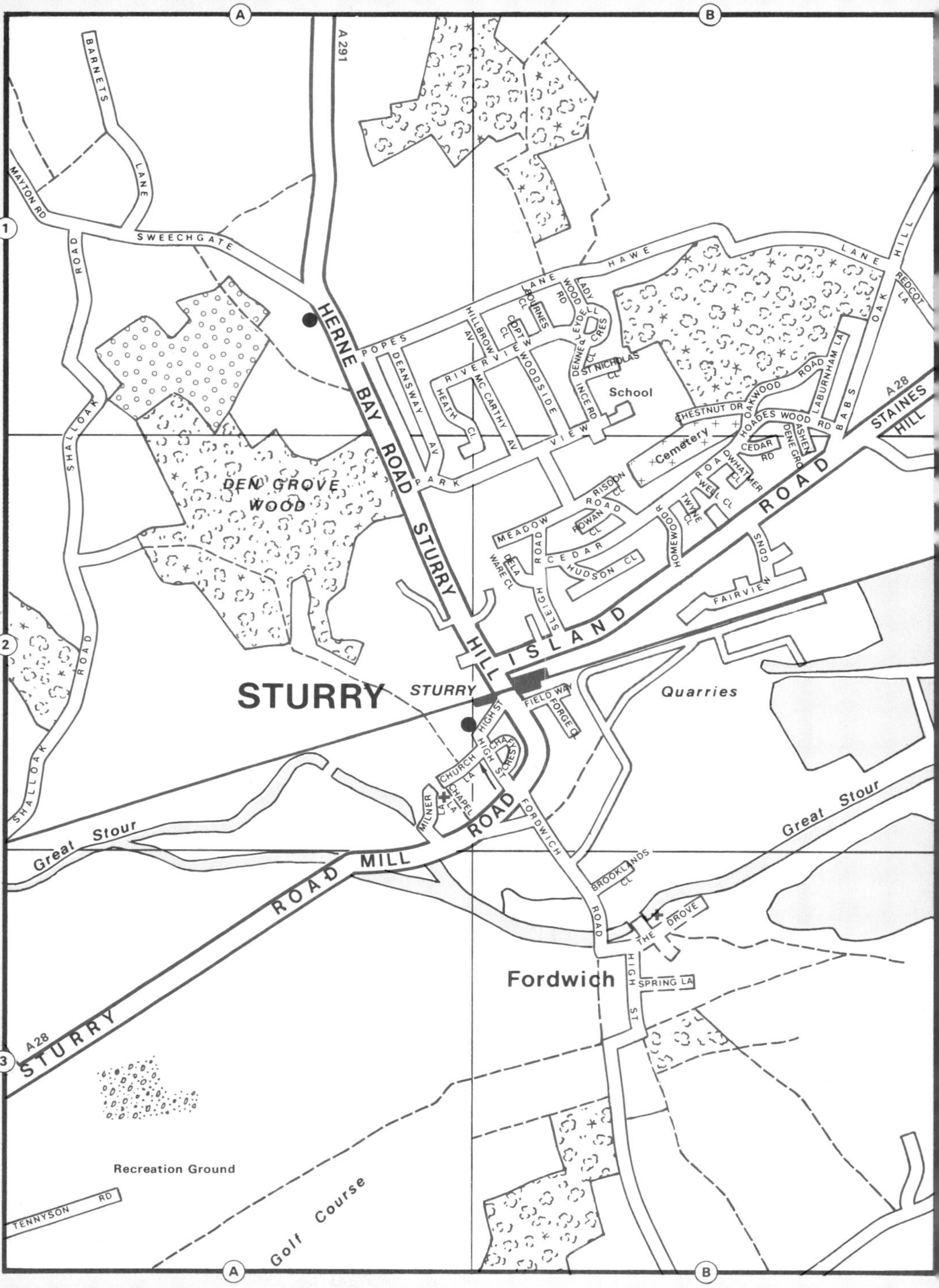

STURRY
Road Map page 13   Population 5,967
For extended coverage of this area—see Estate Publications RED BOOK—Thanet/Canterbury
Reproduction prohibited without prior permission   © Estate Publications

Scale 4 inches to 1 mile

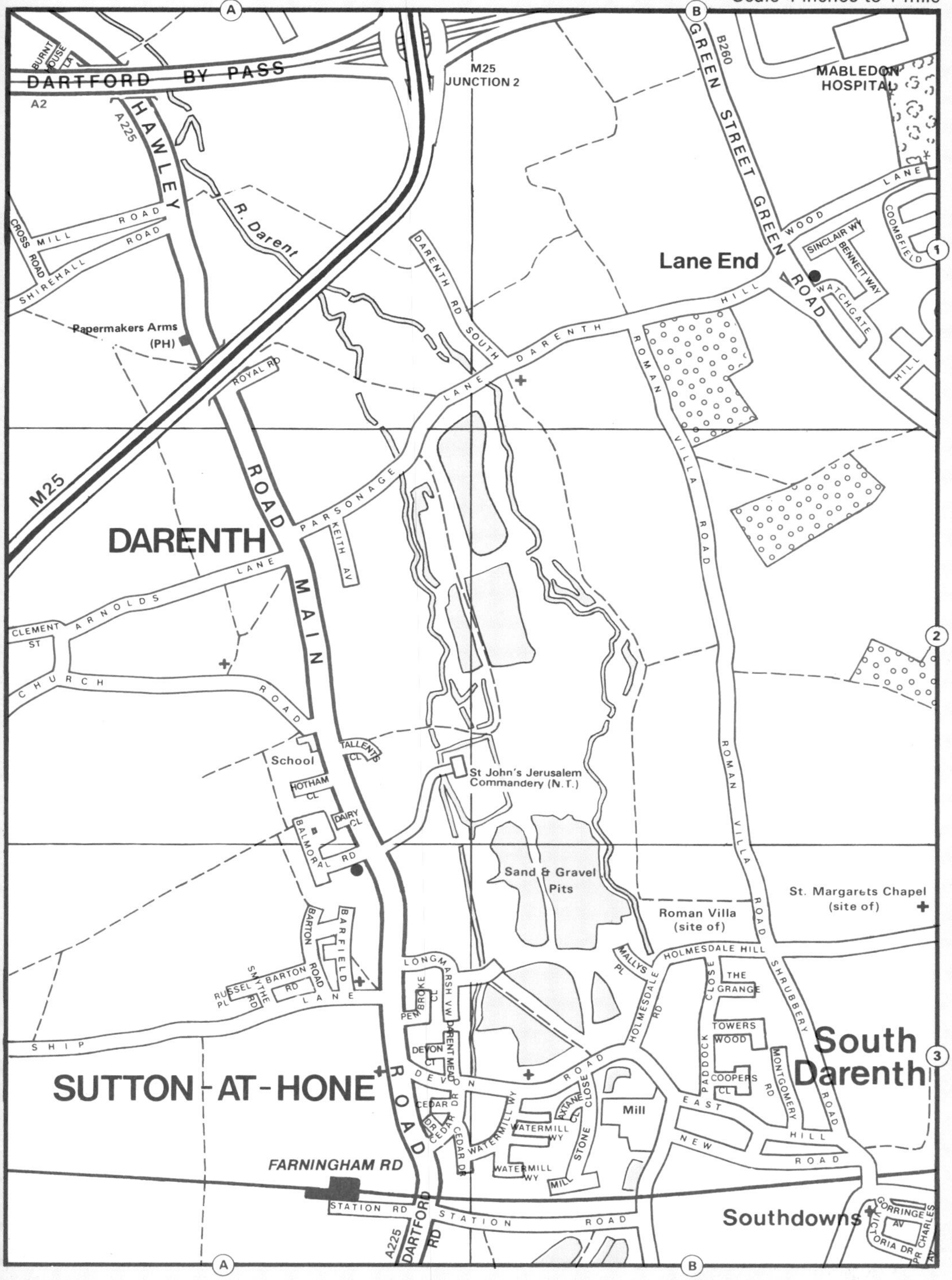

DARENTH   Population 4,942   SUTTON-AT-HONE   Population 3,956
Road Map page 10

SWANLEY

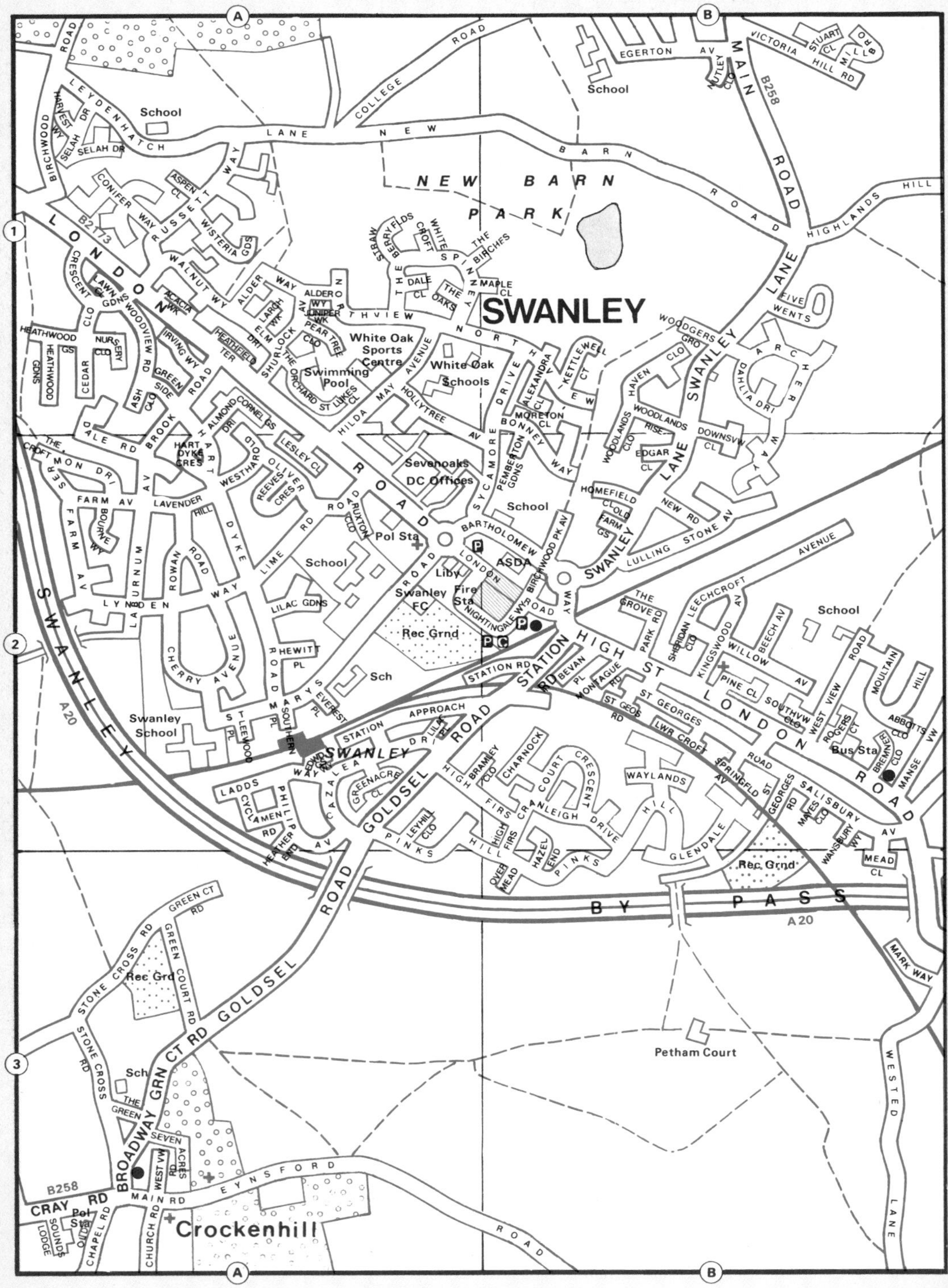

SWANLEY
Road Map page 10   Population 20,945
For extended coverage of this area—see Estate Publications RED BOOK—Gravesend/Dartford
Reproduction prohibited without prior permission   © Estate Publications

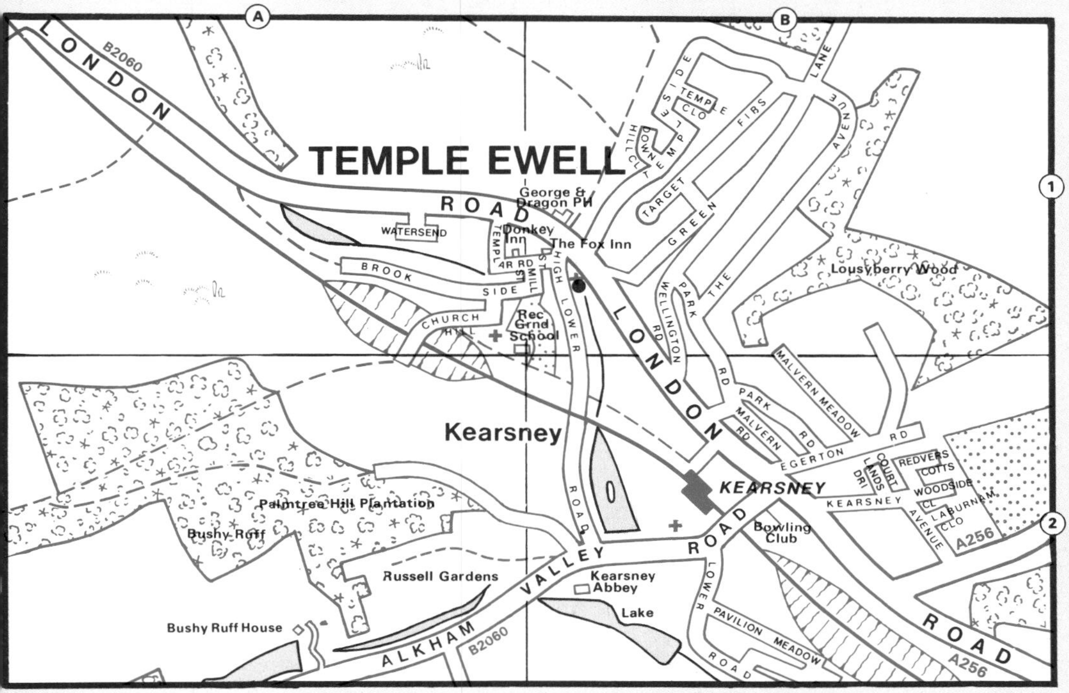

TEMPLE EWELL
Road Map page 13   Population 1,521

For extended coverage of this area—see
Estate Publications RED BOOK—Folkestone/Dover

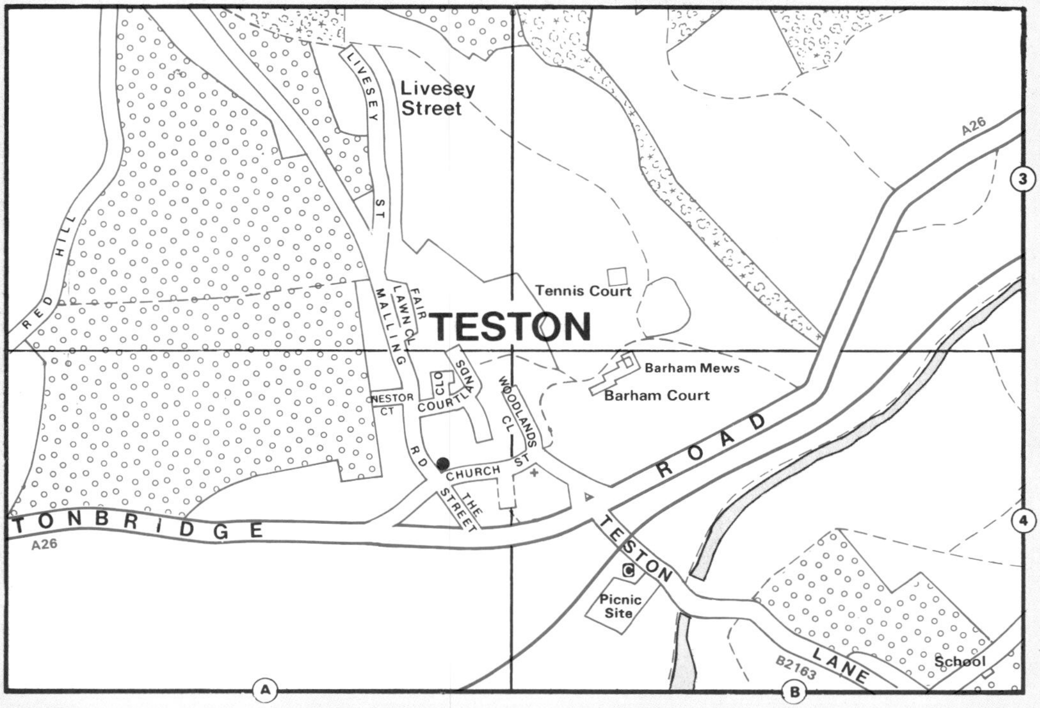

TESTON
Road Map page 11
Reproduction prohibited without prior permission

TESTON

Scale 4 inches to 1 mile

© Estate Publications

Scale 4 inches to 1 mile

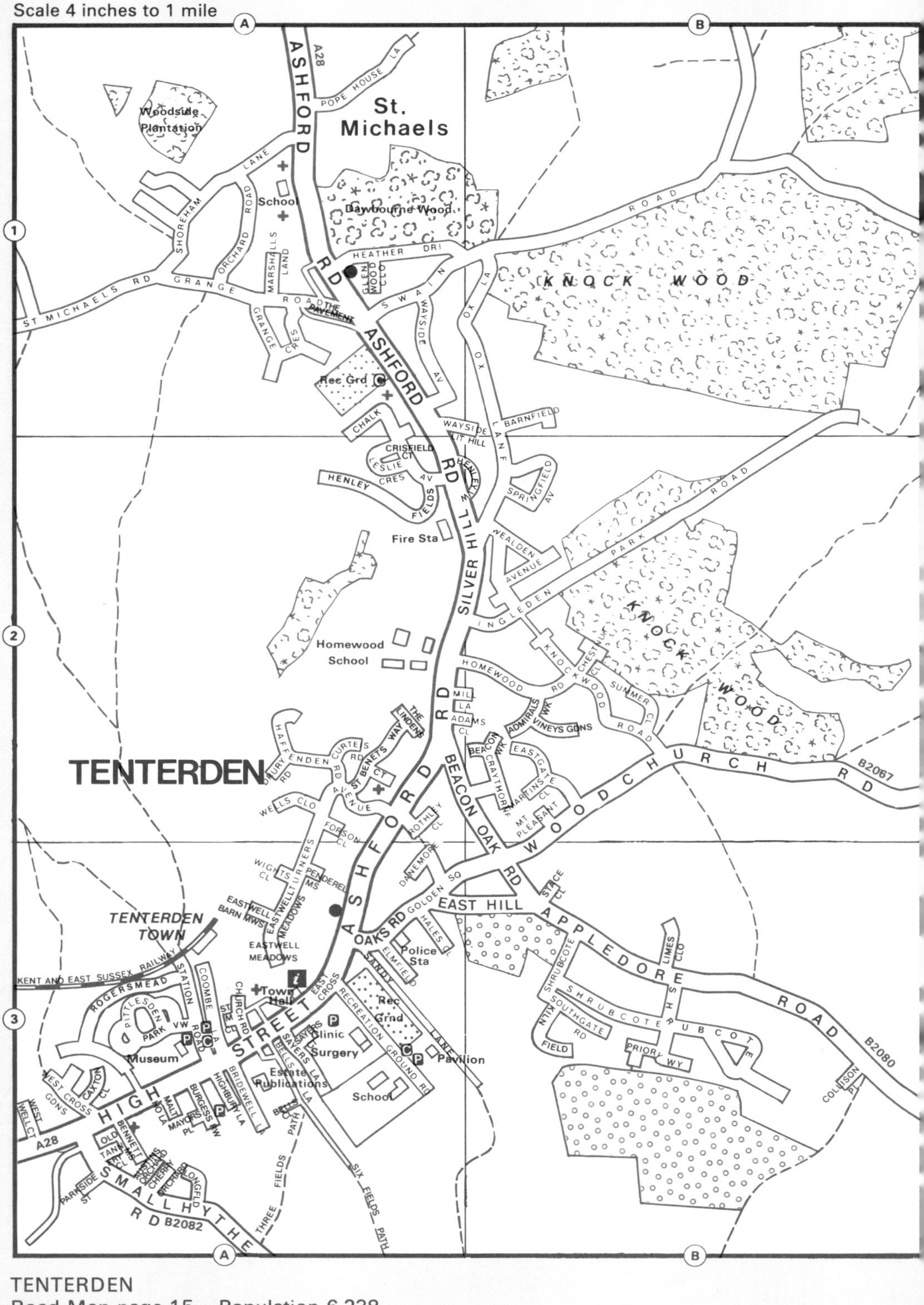

TENTERDEN

Road Map page 15   Population 6,238

For extended coverage of this area — see Estate Publications RED BOOK—Ashford

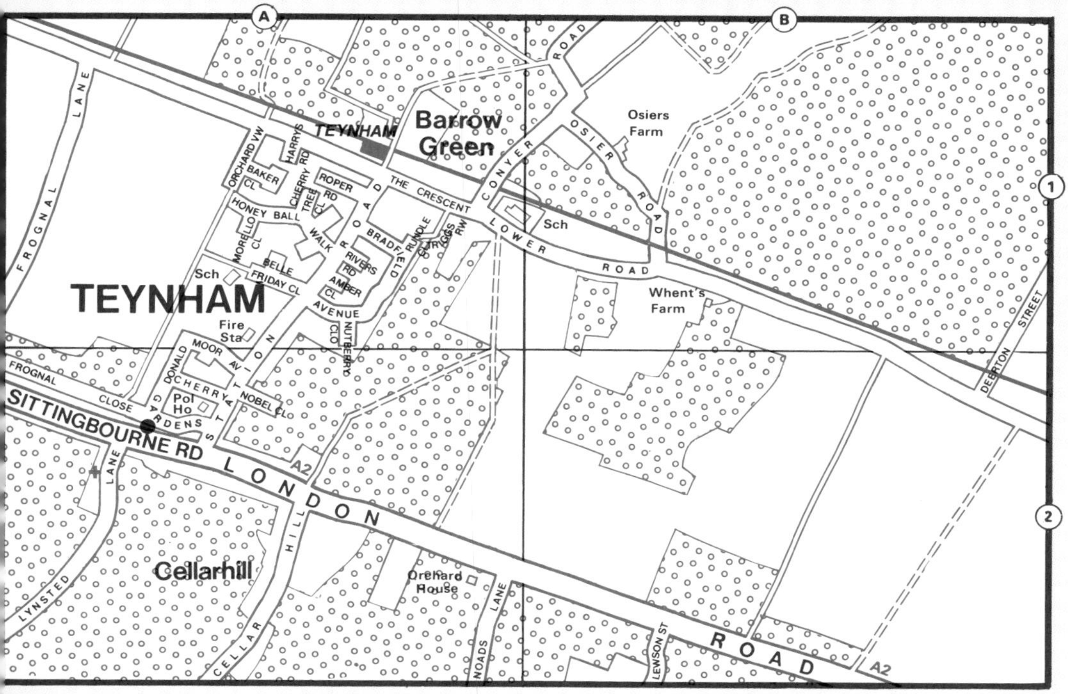

TEYNHAM
Road Map page 12   Population 3,211

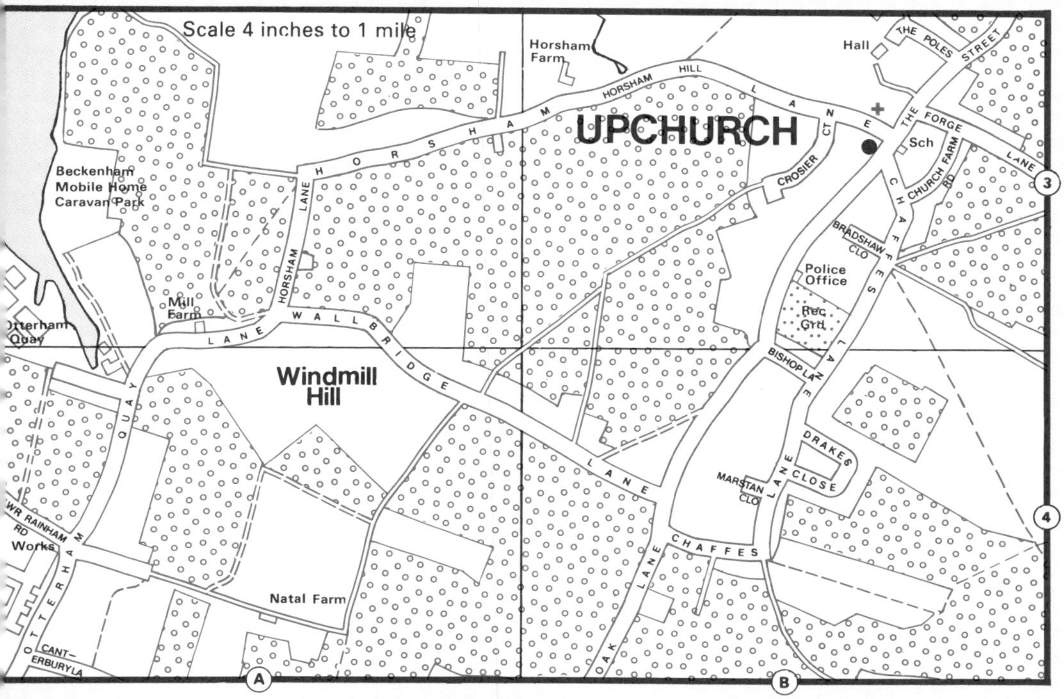

UPCHURCH

Population 1,881

TONBRIDGE

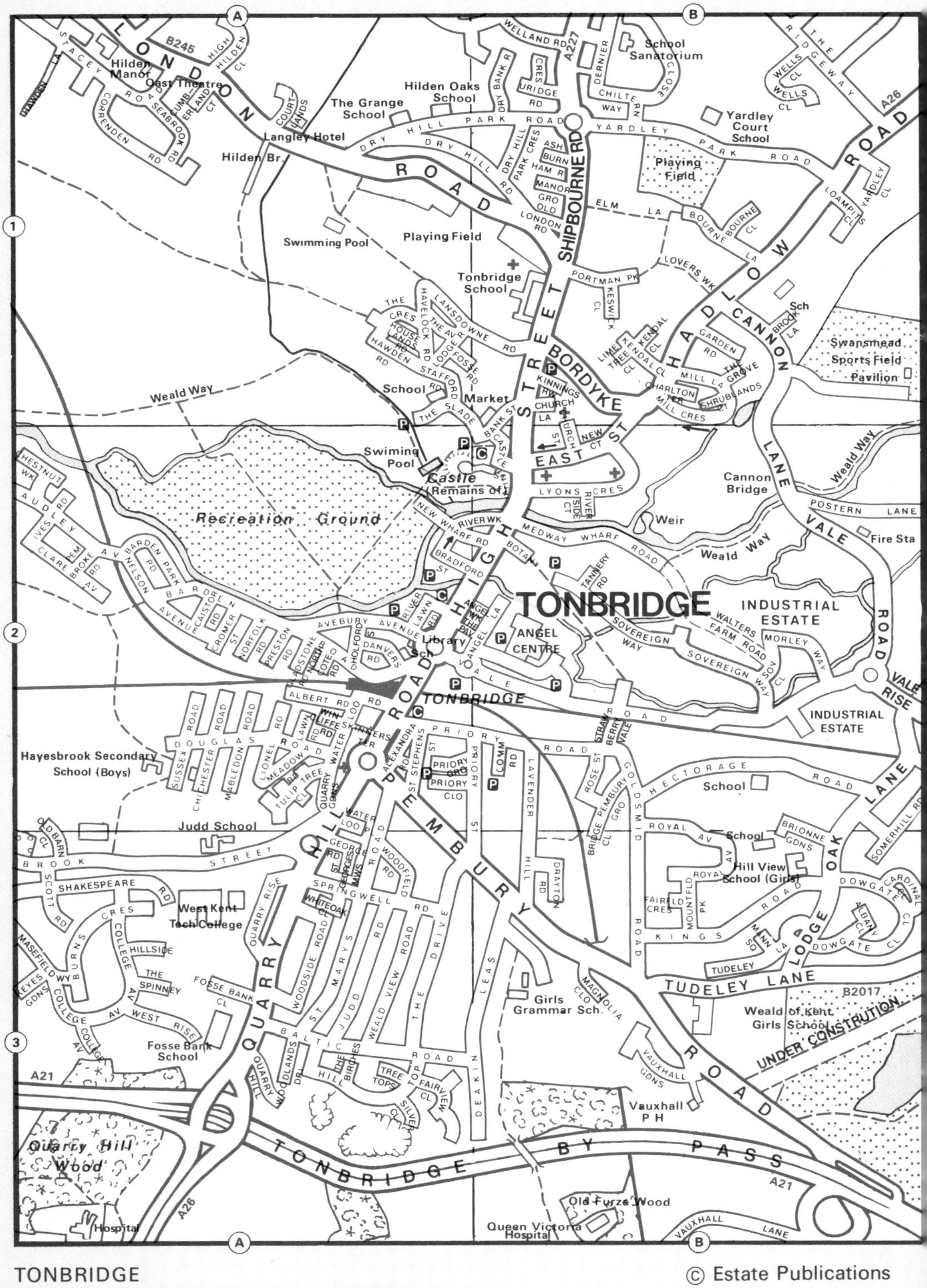

TONBRIDGE
Road Map page 14   Population 30,530
For extended coverage of this area— see Estate Publications RED BOOK—Tunbridge Wells/
Tonbridge
Reproduction prohibited without prior permission

ⓒ Estate Publications

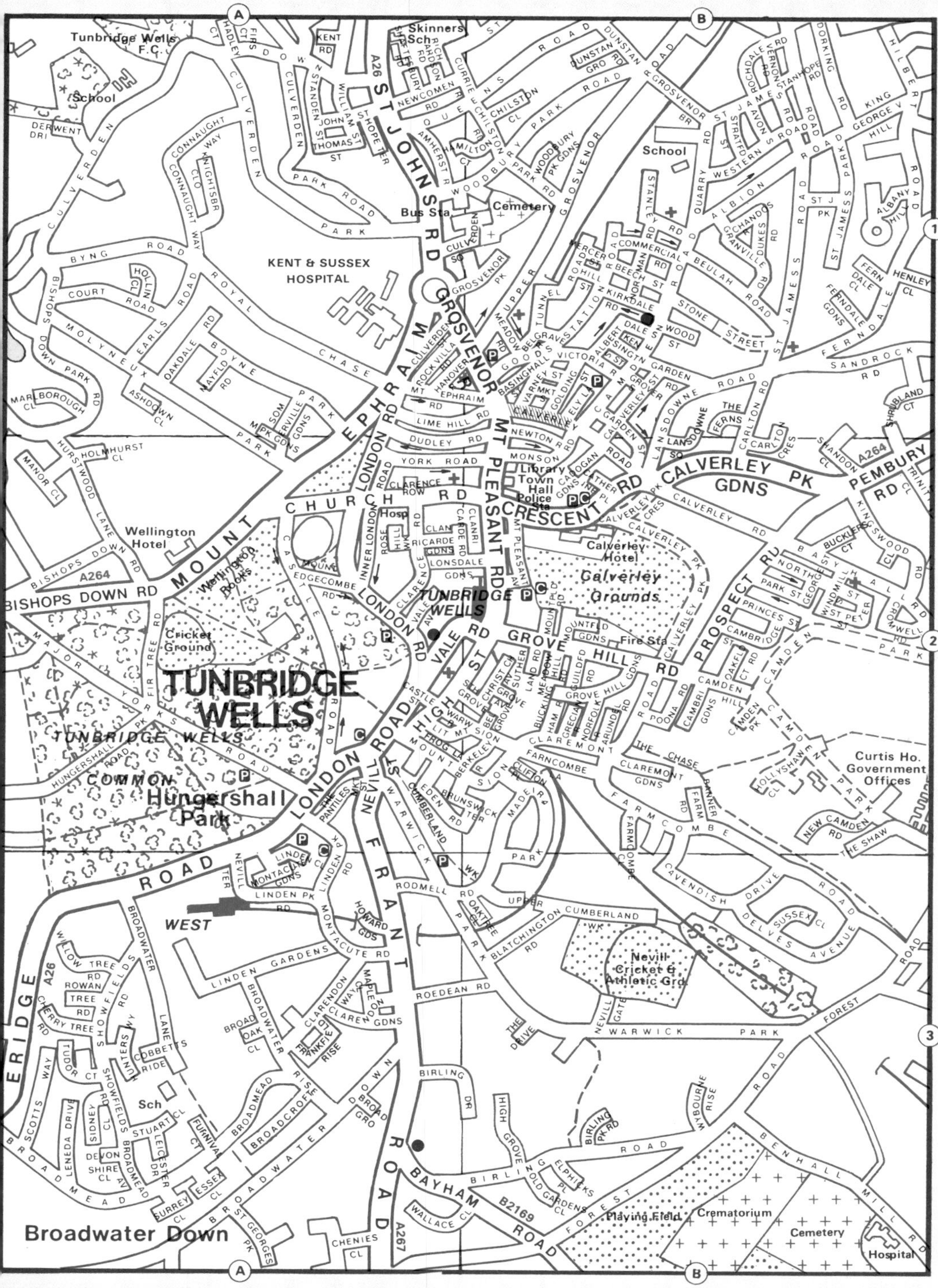

TUNBRIDGE WELLS
Road Map page 14   Population 44,992
For extended coverage of this area— see Estate Publications RED BOOK— Tunbridge Wells/
Reproduction prohibited without prior permission   Tonbridge

© Estate Publications

WALMER

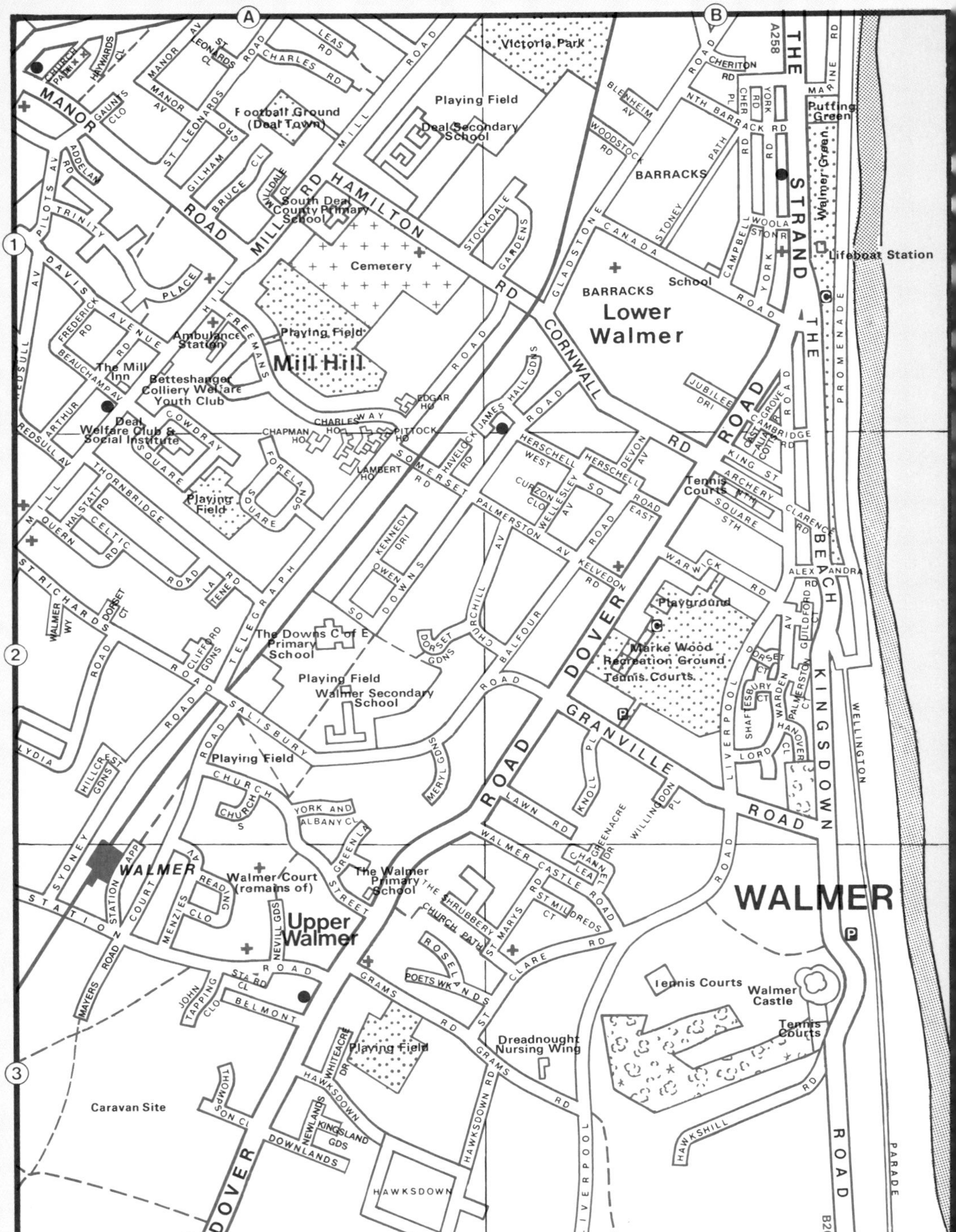

Victoria Park

Football Ground (Deal Town)

Playing Field

Deal Secondary School

MANOR ROAD

South Deal County Primary School

Cemetery

Ambulance Station

Playing Field

The Mill Inn

Betteshanger Colliery Welfare Youth Club

Deal Welfare Club & Social Institute

Mill Hill

Playing Field

RD HAMILTON RD

CORNWALL RD

BARRACKS

Lower Walmer

BARRACKS

School

Tennis Courts

Putting Green

THE STRAND

Lifeboat Station

The Downs C of E Primary School

Playing Field

Walmer Secondary School

Playing Field

Church

York and Albany Cl.

WALMER

Walmer Court (remains of)

The Walmer Primary School

Upper Walmer

DOVER ROAD

Playing Field

Caravan Site

HAWKSDOWN

Playground

Marke Wood Recreation Ground Tennis Courts

GRANVILLE ROAD

DOVER ROAD

KINGSDOWN ROAD

Dreadnought Nursing Wing

Tennis Courts

Walmer Castle

Tennis Courts

WALMER

A258

B2057

WALMER
Road Map page 13

For extended coverage of this area—see Estate Publications RED BOOK—Folkestone/Dover

© Estate Publications

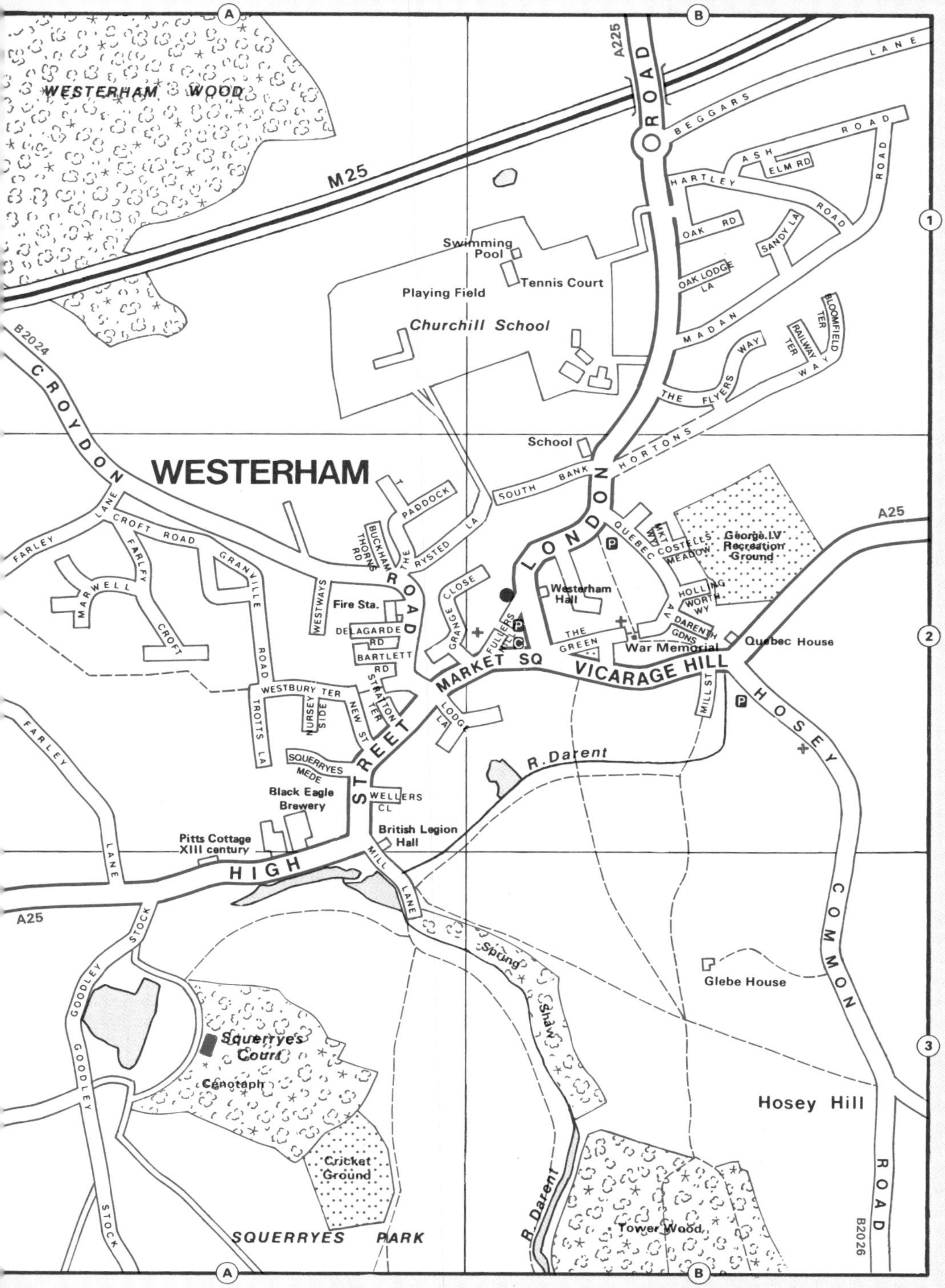

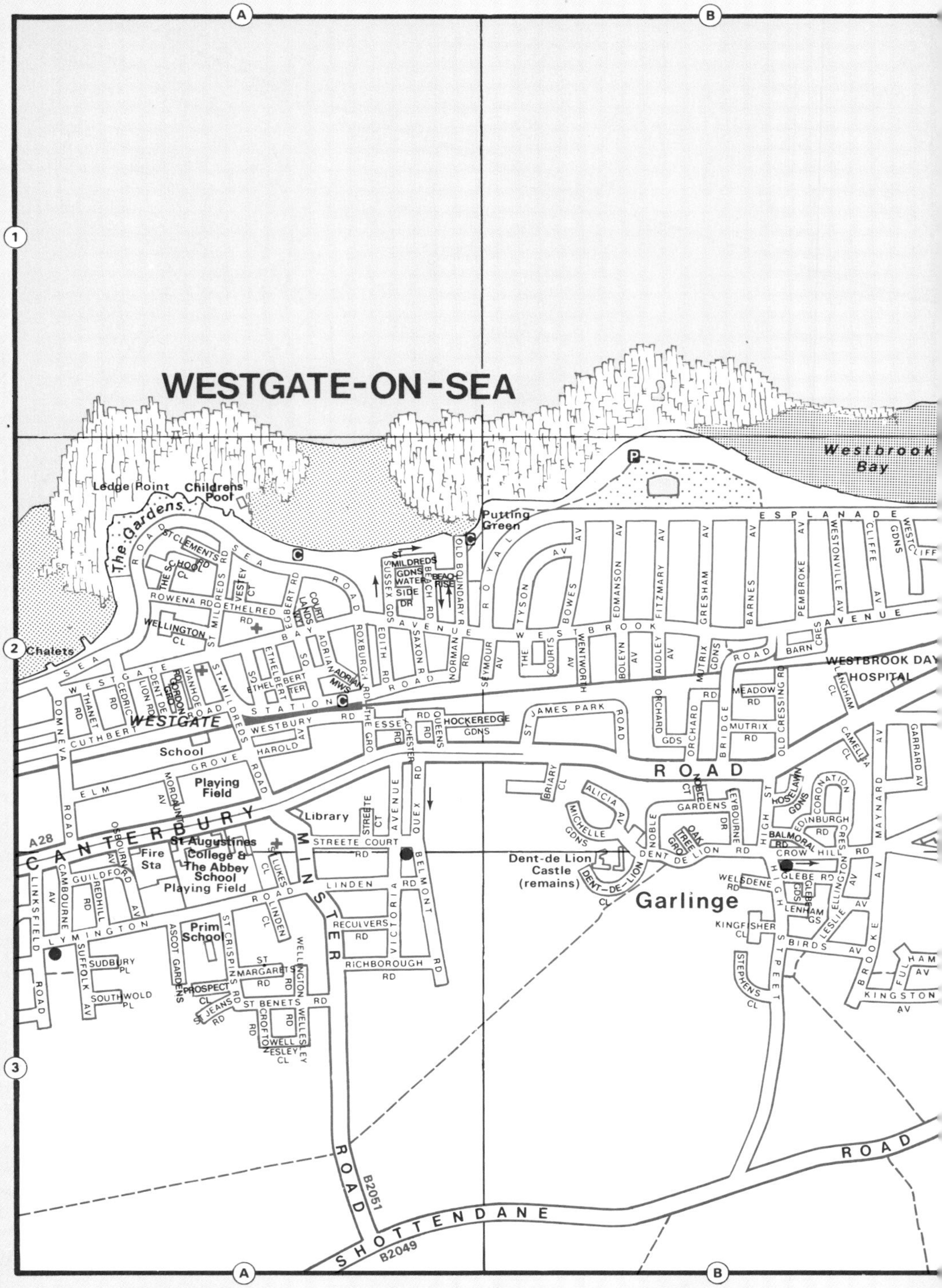

WESTGATE-ON-SEA

For extended coverage of this area—see Estate Publications RED BOOK—Thanet/Canterbury

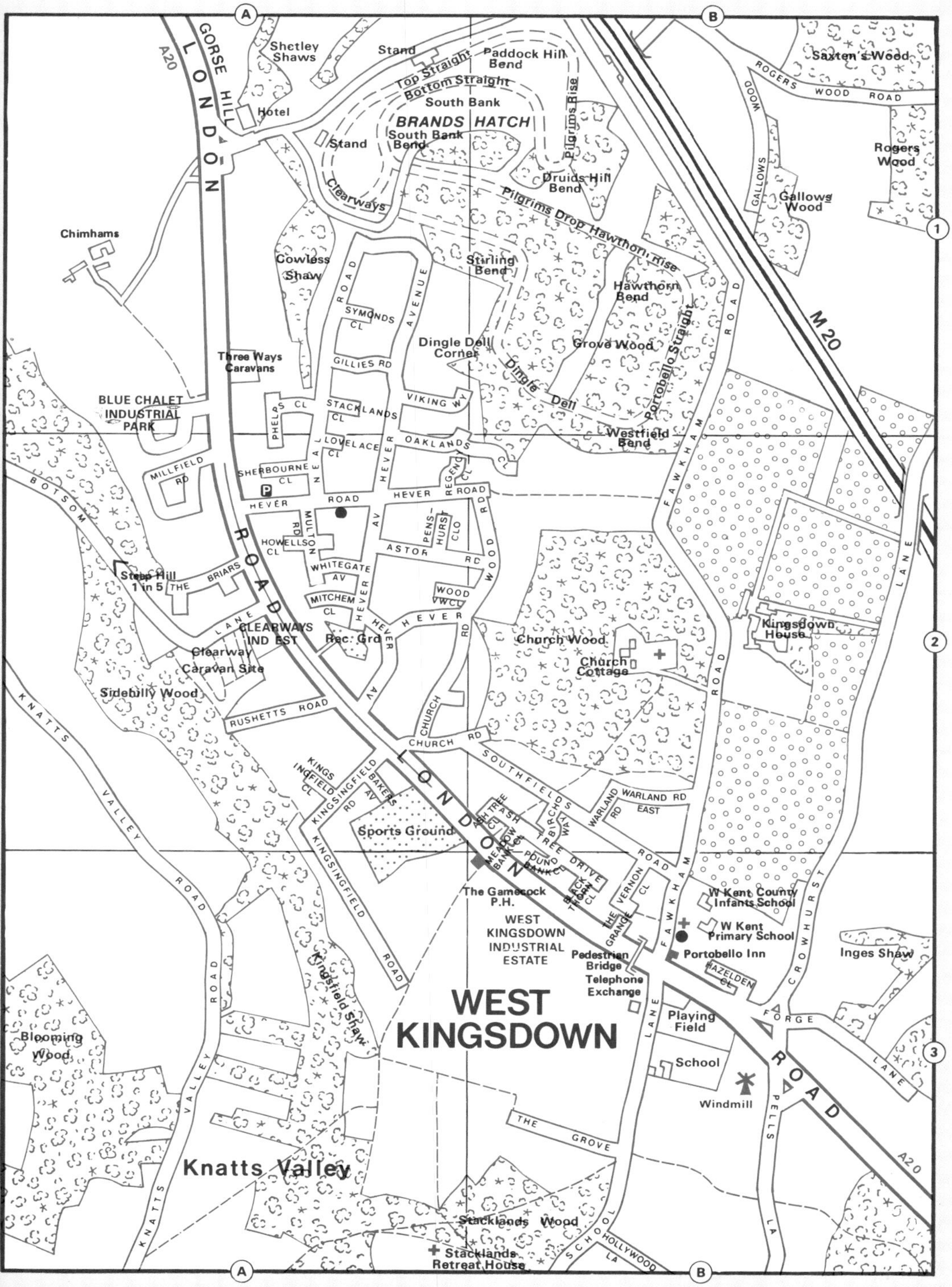

WEST KINGSDOWN

Road Map page 10   Population 4,909

For extended coverage of this area   —   see Estate Publications RED BOOK—Sevenoaks

Reproduction prohibited without prior permission

© Estate Publications

WEST MALLING

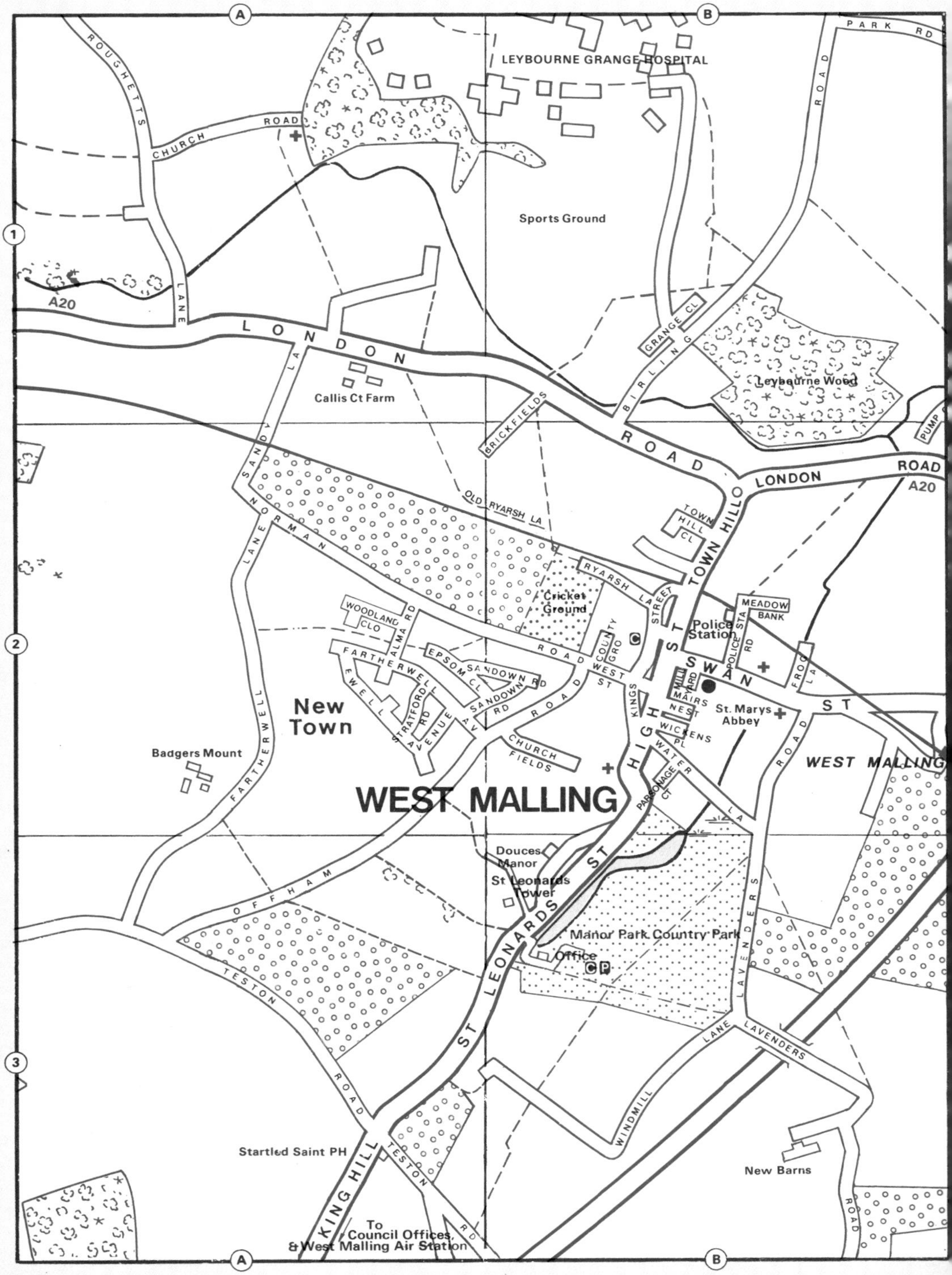

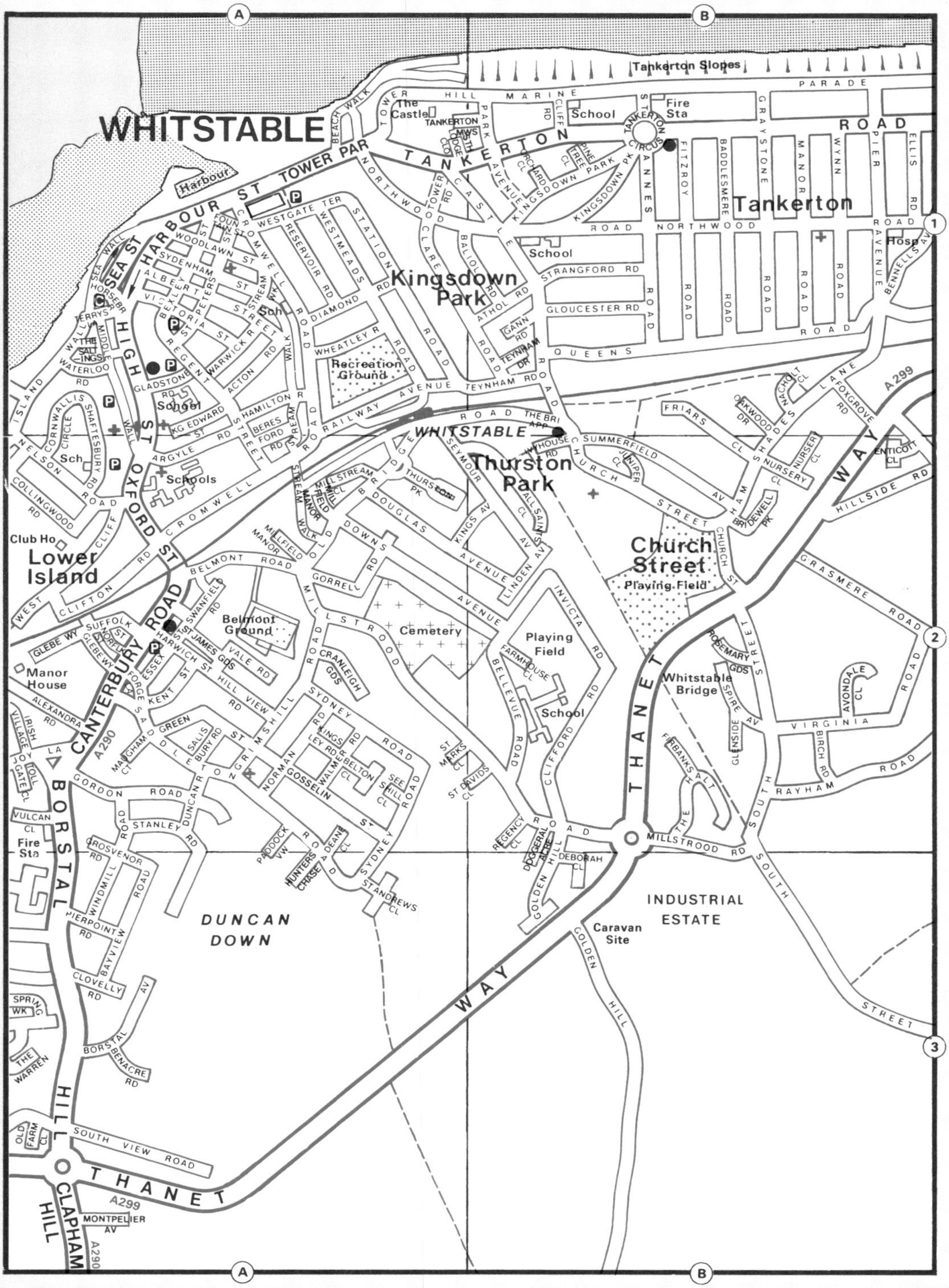

WHITSTABLE
Road Map page 12   Population 27,287
For extended coverage of this area—see Estate Publications RED BOOK—Thanet/Canterbury

WOODCHURCH

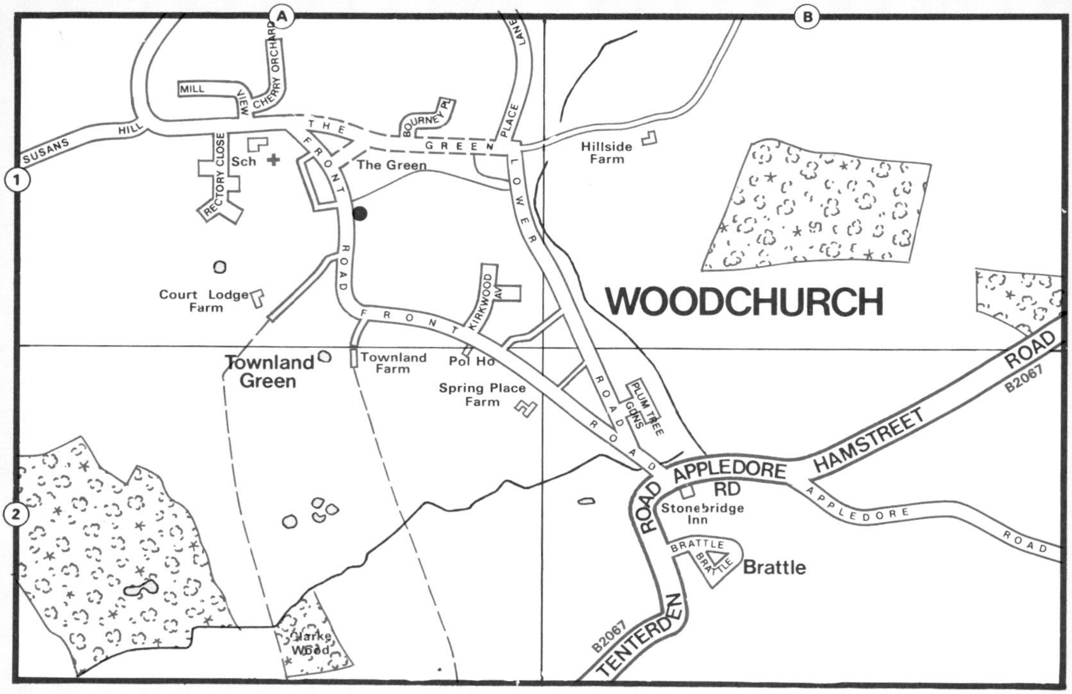

WOODCHURCH
Road Map page 15   Population 1,662

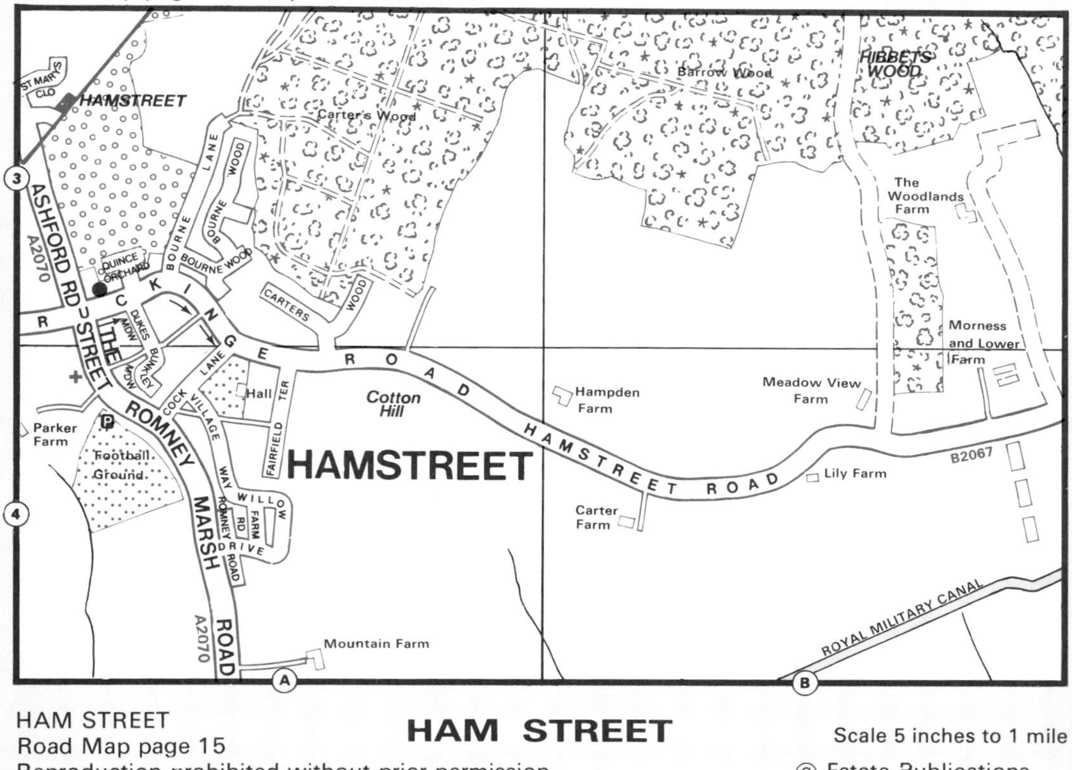

HAM STREET

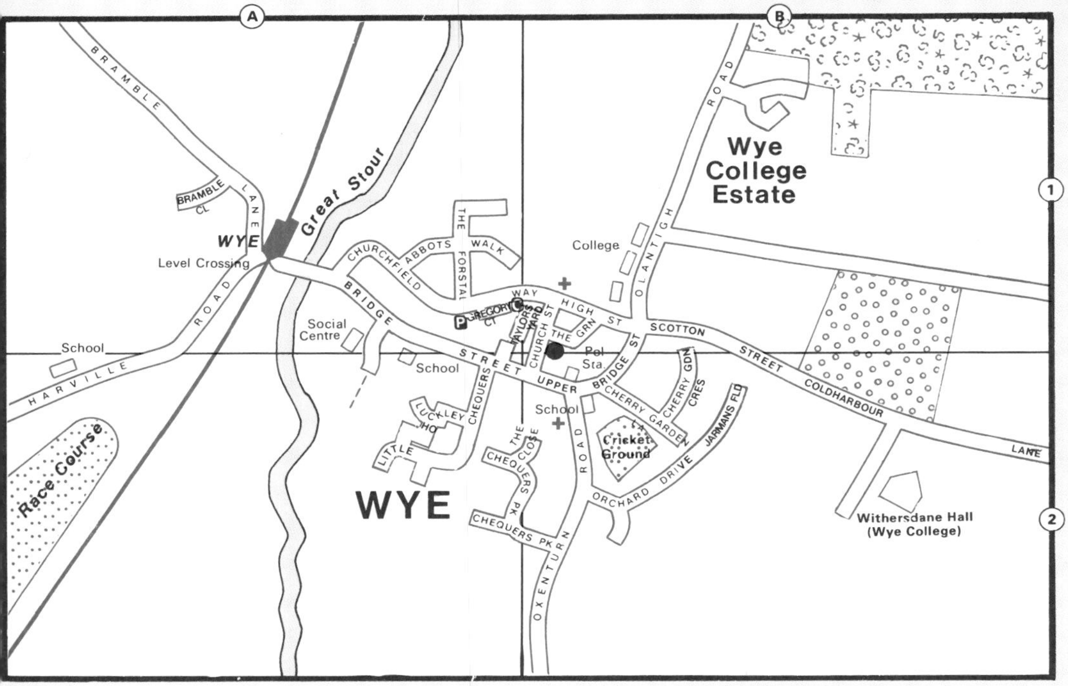

WYE
Road Map page 12   Population 2,000

For extended coverage of this area —
see Estate Publications RED BOOK—Ashford

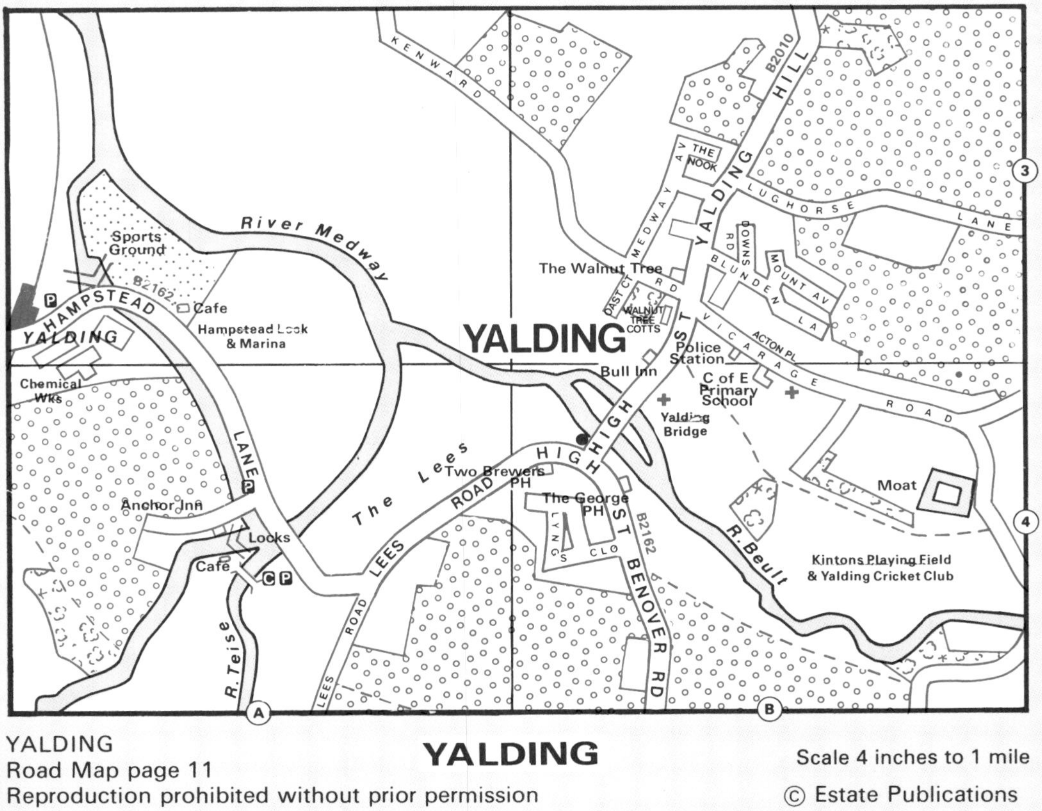

YALDING
Road Map page 11

YALDING

Scale 4 inches to 1 mile

INDEX TO STREETS

Trapfield Clo	19 C1	Ingoldsby Rd	21 A2	Tyler Hill Rd	20 A4	Patrixbourne Rd	23 A2

Trapfield Clo	19 C1
Trapfield La	19 C1
Tydeman Rd	19 B2
Ufton Clo	19 A2
Vintners Way	19 A1
Wagoners Clo	19 A1
Ware St	19 B1
Water La	19 C1
Weavering St	19 A1
Wheatfields	19 A1
Whiteheads La	19 C1
Willington St	19 A1
Wingrove Dri	19 A1
Winifred Rd	19 B1
Wytherling Clo	19 A1
Yeoman La	19 B1
Yeoman Way	19 B2

BIDDENDEN

Chaulkhurst	20 A2
Chaulkhurst Clo	20 A1
Cheeselands	20 A1
Church Vw	20 A1
Cloth Hall Gdns	20 B1
Glebelands	20 B2
High St	20 B2
North St	20 B1
Shuttle Clo	20 B1
Sissinghurst Rd	20 A2
Spinners Clo	20 B1
Tenterden Rd	20 B2
The Meadows	20 B1
The Weavers	20 B2
Townland Clo	20 B1

BIRCHINGTON

Acol Hill	21 B3
Albion Rd	21 B2
Alexandra Rd	21 B3
Alfred Rd	21 A1
Allison Clo	21 B1
Alpha Rd	21 B2
Anna Park	21 A1
Anne Clo	21 B2
Anvil Clo	21 B2
Arthur Rd	21 A1
Barrington Cres	21 B2
Barrows Clo	21 B2
Beach Av	21 A2
Berkeley Rd	21 A1
Beverley Clo	21 A1
Bierce Ct	21 A2
Broadley Av	21 A3
Brunswick Rd	21 B3
Canterbury Rd	21 A3
Chapel Pl	21 B2
Charlesworth Dri	21 B2
Cliff Rd	21 A1
Colemans Stairs	21 B1
Colemans Stairs Rd	21 B1
Constable Rd	21 B1
Conway Clo	21 A2
Cornford Rd	21 B2
Crescent Rd	21 B2
Crispe Rd	21 A3
Cross Rd	21 B1
Cunningham Cres	21 A2
Dallinger Rd	21 A1
Darwin Rd	21 A2
Devon Gdns	21 A2
Dorset Gdns	21 A2
Duncan Dri	21 A2
Eastfield Rd	21 B2
Edenfield	21 B2
Edward Dri	21 B2
Egbert Rd	21 A2
Epple Bay Av	21 B1
Epple Bay Rd	21 A2
Epple Cottages	21 B1
Epple Rd	21 B1
Essex Gdns	21 A2
Farrar Rd	21 B2
Ferndale Ct	21 B2
Gainsborough Rd	21 A1
Gallwey Av	21 A2
Gordon Sq	21 A2
Gore End Clo	21 A2
Green Rd	21 A1
Grenham Bay Rd	21 A2
Grenham Rd	21 A1
Grenville Gdns	21 A1
Harold Rd	21 A1
Hawkhurst Clo	21 B1
Hereford Gdns	21 A2
Hereward Av	21 A1
Hershell Rd	21 A1
Hoser Gdns	21 B2
Hunting Gate	21 A2
Ingle Clo	21 B2

Ingoldsby Rd	21 A2
Kent Gdns	21 A2
King Edward Rd	21 A3
Kings Rd	21 B3
Laming Rd	21 B2
Lancaster Gdns	21 A2
Leslie Rd	21 B1
Lincoln Gdns	21 A2
Linnington Rd	21 B2
Lyell Rd	21 A2
Magnus Ct	21 B1
Manor Dri	21 A2
Manston Rd	21 B3
Marilyn Cres	21 B2
Mellanby Clo	21 B2
Melsetter Clo	21 B2
Mill La	21 A2
Mill Row	21 A2
Minnis Rd	21 A2
Moray Av	21 A1
Nasmyth Rd	21 B1
Neame Rd	21 B2
Nelson Clo	21 A1
Norrie Rd	21 B2
Nottingham Rd	21 A3
Ocean Clo	21 B1
Old Farm Rd	21 A2
Oxney Clo	21 B2
Paddock Rd	21 B2
Park La	21 B2
Park Rd	21 B2
Phillips Rd	21 B2
Powell Cotton Dri	21 B2
Princes Clo	21 A2
Prospect Rd	21 A2
Quex View Rd	21 A2
Reculver Rd	21 A2
Romney Clo	21 B2
Rose Gdns	21 A2
Rossetti Rd	21 A2
Rutland Gdns	21 A2
St Davids Clo	21 B2
St James Ter	21 B1
St James Ter	21 B2
St Magnus Clo	21 B1
St Mildreds Av	21 A2
Sandles Rd	21 A2
Sea Rd	21 B1
Sea View Av	21 A1
Sea View Rd	21 A1
Seamark Rd	21 A3
Semaphore Rd	21 A1
Sewell Clo	21 B2
Shakespeare Rd	21 A1
Sheppey Clo	21 B2
Sherwood Rd	21 A3
Silver Av	21 B3
Smugglers Wy	21 B1
Spencer Rd	21 A1
Station App	21 B2
Station Rd	21 B2
Stone Barn Av	21 B2
Surrey Gdns	21 A2
Sussex Gdns	21 A2
The Parade	21 A1
The Retreat	21 B1
The Square	21 B2
Tower Bungalows	21 B1
Tudor Clo	21 B2
Walnut Tree Pl	21 B2
Westfield Rd	21 A2
Wilbrough Rd	21 B2
Wilkie Rd	21 B1
Winston Ct	21 B1
Woodford Ct	21 B2
Woodland Av	21 B2
Yew Tree Gdns	21 B2
York Ter	21 B2

BLEAN

Blean Common	20 A3
Blean Hill	20 A4
Bourne Lodge Clo	20 A3
Calais Hill	20 B4
Canterbury Hill	20 B4
Chapel La	20 A3
Chestnut Av	20 A3
Fleet La	20 B3
Giles La	20 B4
Hackington Rd	20 B3
Link Rd	20 B4
Lucketts La	20 A4
Mount Pleasant Clo	20 A4
St Johns Cres	20 B3
St Stephens Hill	20 B4
School La	20 A3
Summer La	20 B4
Sunny Mead	20 B3
Tile Kiln Hill	20 A4
Truman Clo	20 A3

Tyler Hill Rd	20 A4
Vicarage La	20 A4
Westfield	20 A4
Wood Hill	20 B4

BOROUGH GREEN/ WROTHAM

Abbott Rd	22 A3
Annetts Hall	22 A2
Ascot Clo	22 B2
Bancroft Rd	22 A1
Battlefields Rd	22 A1
Beechin Wood Rd	22 B3
Black Horse La	22 B3
Bracken Hill	22 B3
Brockway	22 A3
Bunish La	22 B3
Comp La	22 B3
Conyerd Rd	22 A3
Crouch La	22 A3
Crow Hill	22 A3
Crowhill Rd	22 A2
Crowhurst Rd	22 A3
Dene Lodge Clo	22 A2
Dryland Rd	22 A3
Eaglestone Clo	22 A2
Fairfield Rd	22 A2
Goodworth Rd	22 A1
Grange Rd	22 B3
Greenlands	22 B3
Griggs Way	22 A3
Harrison Rd	22 A3
High St, Borough Green	22 A3
High St, Wrotham	22 A1
Hill View	22 A3
Hill View Clo	22 A3
Hunts Farm Clo	22 A3
Kemsing Rd	22 A1
Lendon Rd	22 A3
Linfield Rd	22 A2
London Rd	22 A1
Long Mill La	22 B3
McDermott Rd	22 A3
Maidstone Rd	22 A3
Minters Orchard	22 B3
Monkton Rd	22 A3
Mountfield	22 A3
Nepicar La	22 B1
New Walk	22 A1
Normanhurst	22 A3
Pine View	22 B2
Platt Common	22 B2
Quarry Hill Rd	22 A3
Randall Hill Rd	22 A1
Riggs Wk	22 A1
Rock Rd	22 A3
St Georges Ct	22 A1
St Marys Clo	22 B3
St Marys Rd	22 A1
Sandy Ridge	22 A2
Staleys Rd	22 A3
Station Rd	22 A2
The Avenue	22 A2
The Close	22 A2
The Crescent	22 A2
The Landway	22 A3
Thomas Wyatt Way	22 A1
Tilton Rd	22 A3
Tolsey Mede	22 A2
West St	22 A1
Western Rd	22 A2
Whitehill	22 B1
Wrotham By-Pass	22 A1
Wrotham Rd	22 A1
Wyatt Clo	22 A3
Wye Rd	22 A2

BRIDGE

Beech Hill	23 B2
Bekesbourne Rd	23 A1
Bourne Vw	23 A2
Bridge By-Pass	23 A1
Bridge Down	23 B2
Bridge Hill	23 A2
Bridgeford Way	23 A2
Churchill Clo	23 A1
Conyngham La	23 A1
Dering Clo	23 A1
Dering Rd	23 A1
Filmer Rd	23 A1
Ford Clo	23 A2
Green Ct	23 A2
High St	23 A1
Higham La	23 B2
Keepers Hill	23 B1
Meadow Clo	23 A2
Mill La	23 A2
Mill Ter	23 A2

Patrixbourne Rd	23 A2
Pett Hill	23 A1
Riverside Clo	23 A1
Saxon Rd	23 A1
Station Rd	23 A1
The New Clo	23 A1
Union Rd	23 A1
Western AV	23 A2
Windmill Clo	23 A2

BROADSTAIRS

Afghan Rd	24 A1
Albert Rd	24 A1
Albion Rd	24 A2
Albion St	24 B2
Alderney Gdns	24 A2
Alexandra Rd	24 B2
Annes Rd	24 B1
Approach Rd	24 A3
Astor Rd	24 A1
Bairdsley Clo	24 A2
Balliol Rd	24 A1
Beacon Rd	24 A2
Beaconsfield Gdns	24 A2
Bedford Ct	24 B2
Belmont Rd	24 B2
Belvedere Rd	24 B3
Birds Hill	24 A2
Bishops Av	24 B1
Bracken Ct	24 A2
Bradstow Way	24 A2
Brassey Av	24 A3
Briars Walk	24 A3
Broadstairs Rd	24 A2
Bromstone Mews	24 A3
Bromstone Rd	24 A3
Buckingham Rd	24 B3
Caernarvon Gdns	24 B2
Callis Court Rd	24 A2
Camden Rd	24 A1
Carlton Av	24 B2
Caroline Cres	24 A2
Castle Av	24 B1
Catherine Way	24 A2
Cecilia Gro	24 A1
Cedar Clo	24 A1
Chandos Rd	24 B3
Chandos Sq	24 B3
Chaucer Rd	24 A3
Cheviot Ct	24 B2
Church Rd	24 B2
Church Sq	24 B2
Church St	24 A2
Claire Ct	24 B2
Clarendon Mews	24 A2
Clarendon Rd	24 B3
Cliff Promenade	24 B1
Cliff Rd	24 B1
Collingwood Clo	24 A3
Convent Rd	24 A1
Cornwallis Gdns	24 B1
Coronation Clo	24 A1
Crampton Ct	24 A2
Crawford Rd	24 A2
Crescent Rd	24 B1
Crofts Pl	24 B2
Crow Hill	24 B2
Cumberland Av	24 B2
Dalmaney Gdns	24 B2
Darnley Clo	24 A3
Davids Clo	24 B3
Devonshire Ter	24 B2
Dickens Rd	24 B2
Dickens Walk	24 B2
Dorcas Gdns	24 A1
Douglas Clo	24 A2
Dumpton Park Dri	24 A3
Dundonald Rd	24 B3
East Cliff Prom	24 B2
Eastern Esplanade	24 B2
Edge End Rd	24 A3
Elmwood Av	24 B1
Elmwood Clo	24 A1
Elmdon Pl	24 B2
Ethel Rd	24 A2
Fair St	24 A3
Fairacre	24 A3
Fairfield Rd	24 A3
Fern Ct	24 B2
Fig Tree Rd	24 A1
Fordoun Rd	24 A2
Fordwich Gro	24 A1
Fort Rd	24 B2
Fosters Av	24 A1
Francis Rd	24 B1
Gladstone Rd	24 A3
Gloucester Av	24 A3
Grafton Rd	24 A1
Grange Rd	24 A1
Grange Way	24 A3

Grant Clo	24 A2
Granville Av	24 A3
Granville Rd	24 B3
Green La	24 A2
Greville Way	24 A3
Grosvenor Rd	24 A3
Guy Clo	24 B1
Harbour St	24 B2
Harrow Dene	24 A2
Harrow Dri	24 A2
High St, Broadstairs	24 B2
High St, St Peters	24 A2
Hildersham Clo	24 A2
Holm Oak Gdns	24 A3
Howard Rd	24 A3
Hubert Way	24 A1
Hugin Av	24 A1
Inverness Ter	24 B3
Joanns Ct	24 B3
Julie Clo	24 A1
Kendal Rise	24 A2
King Edwards Av	24 B3
Kings Av	24 B2
Knights Av	24 B2
Lanthorne Rd	24 A1
Lauriston Mount	24 A2
Lawn Rd	24 B2
Laxing Av	24 B1
Leatt Clo	24 A3
Lerryn Gdns	24 A1
Leybourn Rd	24 B3
Linden Av	24 B2
Lindenthorpe Rd	24 A2
Link Rd	24 A1
Linley Rd	24 A1
Livingstone Rd	24 A2
Lloyd Rd	24 A2
Luton Av	24 A3
Luton Ct	24 A3
Lyndhurst Rd	24 B2
Magdala Rd	24 A2
Magdalen Ct	24 B2
Manor Rd	24 A3
Marlborough Clo	24 A3
Marshall Cres	24 A3
Masons Rise	24 A2
Maxine Gdns	24 A2
Mayville Rd	24 A1
Millfield	24 A2
Mockett Dri	24 A1
Napier Rd	24 A2
Nash Gdns	24 B2
Nelson Pl	24 B2
Norman Rd	24 A2
North Foreland Av	24 B1
North Foreland Rd	24 B1
Northcliffe Gdns	24 A1
Northdown Rd	24 A1
Oaklands Av	24 A2
Old Green Rd	24 A1
Osborne Rd	24 A3
Oscar Rd	24 B3
Palmerston Av	24 B3
Park Av	24 A3
Park Chase	24 A3
Park Gate	24 A3
Park Rd	24 B2
Parkland Ct	24 A2
Percy Rd	24 A2
Pier App	24 B2
Pierremont Av	24 B2
Poplar Rd	24 A2
Prince Andrew Rd	24 A1
Prince Charles Rd	24 A1
Princess Anne Rd	24 A1
Priory Clo	24 A3
Promenade	24 B3
Prospect Pl	24 B2
Prospect Rd	24 B2
Queens Av	24 B2
Queens Gdns	24 B3
Queens Rd	24 B3
Radley Clo	24 B2
Raglan Pl	24 B3
Ramsgate Rd	24 A3
Ranelagh Gro	24 A2
Reading St	24 A1
Reading Street Rd	24 A1
Rectory Rd	24 B2
Repton Clo	24 A2
Rhodes Gdns	24 A1
Rosemary Av	24 A3
Rosemary Gdns	24 A3
Rosetower Ct	24 A1
Rugby Clo	24 A2
St Christopher Grn	24 A2
St Georges Rd	24 B3
St James Av	24 A2
St Marys Rd	24 B2
St Mildreds Av	24 A3
St Peters Ct	24 A2

St Peters Park Rd	24 A2
Salisbury Av	24 A3
Salts Dri	24 A2
Sanctuary Clo	24 A3
Sea App	24 B3
Sea View Rd	24 B2
Seafield Rd	24 A3
Seapoint Rd	24 B3
Selwyn Dri	24 A2
Shutter Rd	24 B2
Sowell St	24 A2
Speke Rd	24 A2
Staines Pl	24 B2
Stanley Pl	24 B2
Stanley Rd	24 A1
Stephen Clo	24 B3
Sterling Clo	24 A2
Stone Gdns	24 B2
Stone Rd	24 B2
Swinburne Av	24 A3
Thanet Clo	24 B2
Thanet Place Gdns	24 B1
Thanet Rd	24 B2
The Banks	24 A2
The Oaks	24 A1
The Paddocks	24 A1
The Parade	24 B3
The Pathway	24 B2
The Promenade	24 B3
The Ridgeway	24 A3
The Vale	24 A3
Tina Gdns	24 B1
Tippledore La	24 A2
Trinity Sq	24 A1
Tunis Row	24 B2
Union Sq	24 B2
Upper Approach Rd	24 B3
Upton Rd	24 A2
Vale Rd	24 A3
Vere Rd	24 B2
Victoria Par	24 B3
Victoria Rd	24 A2
Vine Clo	24 A3
Waldron Rd	24 B3
Wallace Way	24 A3
Walmsley Rd	24 A2
Wardour Clo	24 B2
Warren Dri	24 A2
Wayne Clo	24 A2
Wellesley Clo	24 A3
West Cliff Av	24 B3
West Cliff Prom	24 B3
West Cliff Rd	24 B3
Western Esplanade	24 B3
Westover Gdns	24 A1
Westover Rd	24 A1
Wilkes Rd	24 A3
Wings Clo	24 B2
Wrotham Av	24 B3
Wrotham Cres	24 B3
Wrotham Rd	24 B3
Yarrow Clo	24 A3
York Av	24 B3
York St	24 B3

BROMLEY

Aldemary Rd	25 B1
Andace Park Gdns	25 B1
Apollo Dri	25 B1
Appledore Clo	25 A3
Aylesbury Rd	25 A2
Babacombe Rd	25 B1
Barnfield Wood Rd	25 A3
Barnhill Av	25 A3
Beadon Rd	25 B3
Beckenham La	25 A1
Belcroft Clo	25 A1
Benenden Grn	25 A3
Bidborough Clo	25 A3
Blendon Path	25 A1
Blyth Rd	25 A1
Bourn Vale	25 B3
Bracken Hill Clo	25 A1
Bracken Hill La	25 A1
Brambledown Clo	25 A3
Brenchley Clo	25 A3
Broadoaks Way	25 A3
Bromley Av	25 A1
Bromley Cres	25 A2
Bromley Gdns	25 A2
Cambridge Rd	25 B1
Cameron Rd	25 B3
Caygill Clo	25 A2
Celtic Av	25 A2
Chart Clo	25 A1
Cheriton Av	25 A3
Cheveney Walk	25 A2
Chiltern Gdns	25 A2
Church Rd, Bromley	25 A2
Church Rd, Shortland	25 A1

College Rd	25 B1
College Slip	25 A1
Cornford Clo	25 B3
Court St	25 B1
Cranbrook Clo	25 B3
Crescent Rd	25 B1
Cromwell Av	25 B2
Cromwell Clo	25 B2
Culverstone Clo	25 A3
Cumberland Rd	25 A2
Dainton Clo	25 B1
Deep Dale Av	25 A3
Devonshire Sq	25 B2
Durham Av	25 A2
Durham Rd	25 A2
Dykes Way	25 A2
East St	25 B1
Eastry Av	25 A3
Edison Rd	25 A1
Elmfield Park	25 B2
Elmfield Rd	25 B2
Elvington Grn	25 A3
Ethelbert Clo	25 B2
Ethelbert Rd	25 B2
Exmouth Rd	25 B2
Fair Acres	25 B3
Fairfield Rd	25 B1
Farnaby Rd	25 A1
Farwig La	25 A1
Fernwood Clo	25 B2
Fletchers Clo	25 B2
Florence Rd	25 B1
Forde Av	25 B2
Forstal Clo	25 A2
Freelands Gro	25 B1
Freelands Rd	25 B1
Fyfe Av	25 B1
Glanville Rd	25 B2
Glassmill La	25 A2
Glebe Rd	25 B1
Goodhart Way	25 A3
Grasmere Rd	25 A1
Green Clo	25 A2
Gwydyr Rd	25 A2
Hammelton Rd	25 A1
Harleyford	25 B1
Hawes Rd	25 B1
Haxted Rd	25 B1
Hayes Chase	25 A3
Hayes La, Hayes	25 B3
Hayes La,	
W. Wickham	25 A3
Hayes Rd	25 B2
Hayesford Park Dri	25 A3
Hazelmere Way	25 B3
Heath Rise	25 A3
Heathfield Rd	25 A1
Henry St	25 B1
Henville Rd	25 B1
High St	25 A1
High Tor Clo	25 B1
Highfield Dri	25 A2
Highland Rd	25 A1
Hillside Rd	25 A2
Holligrave Rd	25 B1
Hollwood Rd	25 B2
Homefield Rd	25 B1
Homesdale Rd	25 B2
Hope Pk	25 A1
Horseley Rd	25 B1
Howard Rd	25 A1
Hurstfield	25 B3
Iden Clo	25 A2
Kentish Way	25 B2
Kingsleigh Walk	25 A3
Kingswood Av	25 A2
Kingswood Rd	25 A2
Knowlton Grn	25 A3
Lancaster Clo	25 A2
Langdon Rd	25 B2
Lansdowne Rd	25 B1
Letchworth Clo	25 B3
Letchworth Dri	25 B3
Leybourne Clo	25 B3
Link Field	25 A3
London Rd	25 A1
Longfield	25 A1
Love La	25 B2
Lowlands Av	25 B1
Ludlow Clo	25 B2
Madeira Av	25 A1
Madison Gdns	25 A2
Mapleton Clo	25 A3
Marden Av	25 A3
Marina Clo	25 A2
Market Sq	25 A1
Martins Rd	25 A1
Masons Hill	25 B2
Matefield Clo	25 B3
Mays Hill Rd	25 A2
Mead Way	25 A3

Meadow Rd	25 A1
Mereworth Clo	25 A3
Mill Vale	25 A2
Mitchell Way	25 B1
Mooreland Rd	25 A1
Morgan Rd	25 A1
Mount Calm Clo	25 B3
Murray Av	25 B2
Napier	25 B2
New Farm Av	25 B2
Newbury Rd	25 A2
Newman Rd	25 B1
North Rd	25 B1
North St	25 B1
Northside Rd	25 B1
Oakham Dri	25 A3
Oaklands Rd	25 A1
Oakwood Av	25 B2
Palace Gro	25 B1
Palace Rd	25 B1
Palace Vw	25 B1
Park End	25 A1
Park Gro	25 B1
Park Hill Rd	25 A2
Park Pl	25 B1
Park Rd	25 B1
Paxton Rd	25 B1
Penshurst Grn	25 A3
Phillips Way	25 B1
Pickhurst La	25 A3
Pickhurst Park	25 A3
Pinewood Rd	25 B2
Plaistow La	25 B1
Plymouth Rd	25 B1
Prospect Pl	25 B2
Queen Anne Av	25 A2
Queens Mead Rd	25 A2
Queens Rd	25 B1
Rafford Way	25 B2
Ravens Clo	25 A2
Ravensbourne Av	25 A1
Ravensbourne Rd	25 B1
Recreation Rd	25 A1
Ridley Clo	25 A2
Ringers Rd	25 B2
Rochester Av	25 B1
Rodway Rd	25 B1
Roman Hurst Av	25 A3
Roman Hurst Gdns	25 A2
Ronald Rd	25 A2
Rutland Gate	25 A2
St Blaise Av	25 B1
St Marks Rd	25 A2
St Marys Av	25 A2
St Pauls Sq	25 A1
Sandford Rd	25 B2
Scotts Rd	25 B2
Sherman Rd	25 B1
Shortlands Gdns	25 A1
Shortlands Rd	25 A2
Simpsons Rd	25 B2
Siward Rd	25 B2
South Hill Rd	25 B2
South St	25 B1
South Vw	25 B1
Speldhurst Clo	25 A1
Stamford Dri	25 A2
Stanley Rd	25 B2
Stanstead Clo	25 A3
Station Rd,	
Bromley North	25 B1
Station Rd,	
Shortlands	25 A1
Stockwell Clo	25 B2
Stone Rd	25 A3
Streamside Clo	25 B2
Talbot Rd	25 B2
Tall Elms Clo	25 A3
Tavistock Rd	25 A2
Tetty Way	25 B2
Teynham Gro	25 A3
The Avenue	25 A3
The Chase	25 B2
The Crescent	25 A3
The Glen	25 A1
The Laurels	25 A2
The Mall	25 B2
Tiger La	25 B2
Tootswood Rd	25 A3
Top Pk	25 A3
Tweedy Rd	25 A1
Upper Park Rd	25 B1
Vale Cotts	25 B2
Valen Leas	25 A2
Valley Rd	25 A1
Vincent Clo	25 B3
Walters Yard	25 A1
Warner Rd	25 A1
Warren Av	25 A1
Wendover Rd	25 B2
West St	25 B1

Westmoreland Rd 25 A3
Weston Gro 25 A1
Weston Rd 25 A1
Wharton Rd 25 B1
White Hart Slip 25 B1
Widmore Rd 25 B1
Willowtree Walk 25 B1
Winchester Clo 25 A2
Winchester Park 25 A2
Winchester Rd 25 A2
Wolfe Clo 25 B3
Woodlea Dri 25 A3

CANTERBURY

Abbots Barton Walk 26 B3
Abbots Clo 26 B2
Ada Rd 26 A3
Adelaide Pl 26 A2
Albert Rd 26 B2
Albion Pl 26 B2
Alma Pl 26 B1
Alma St 26 B1
Ann Green Walk 26 B1
Arran Mews 26 B1
Artillery Gdns 26 B2
Artillery St 26 B2
Barnshaw Rd 26 A1
Barton Mill Rd 26 B1
Beaconsfield Rd 26 A1
Beer Cart La 26 A2
Best La 26 A2
Beverley Rd 26 A1
Birchwood Path 26 A1
Bishops Way 26 A1
Black Griffin La 26 A2
Blackfriars St 26 B2
Bristol Rd 26 B3
Broad Oak Rd 26 B1
Broad St 26 B2
Brockenhurst Clo 26 A1
Brymore Clo 26 B1
Burgate La 26 B2
Burgate 26 B2
Butchery La 26 B2
Cadnam Clo 26 A1
Calcraft Mews 26 B1
Cambridge Rd 26 A3
Cambridge Way 26 A3
Canterbury By-Pass 26 A3
Canterbury La 26 B2
Castle Row 26 A2
Castle St 26 A2
Cherry Gro 26 B1
Church La 26 A2
Church St 26 A2
Church St, Longport 26 B2
Claremont Pl 26 A3
Clement Clo 26 B1
Clifton Gdns 26 A1
Clyde St 26 B1
Cogan Ter 26 A3
Cold Harbour 26 B1
College Rd 26 B2
Coopers La 26 A3
Cossington Rd 26 B2
Cotton Rd 26 A3
Cow La 26 A3
Cowdrey Pl 26 B3
Craddock Rd 26 B2
Cranborne Walk 26 A1
Cromwell Rd 26 B3
Cross St 26 A2
Crown Gdns 26 A2
Cushman Rd 26 A3
Damerham Clo 26 A1
Dover St 26 B2
Duck La 26 B2
Durham Clo 26 A3
Durnford Clo 26 A1
Durovernum Ct 26 B3
Edgar Rd 26 B2
Edward Rd 26 B2
Elham Rd 26 A3
Ersham Rd 26 B3
Ethelbert Rd 26 B3
Farleigh Rd 26 B1
Forty Acres Rd 26 A1
Foxdown Clo 26 A1
Friary Way 26 A1
Gas St 26 A2
Gillon Mews 26 B1
Glenside Av 26 B1
Gordon Rd 26 A3
Gore Mews 26 B1
Gravel Walk 26 B2
Green Cloth Mews 26 B1
Grove Ter 26 A3
Guildford Rd 26 A3
Guildhall St 26 B2
Hackington Dri 26 A1
Hackington Pl 26 A1
Hackington Ter 26 A1
Hales Dri 26 A1
Hallett Walk 26 B1
Hanover Pl 26 A1
Harcourt Dri 26 A1
Harkness Dri 26 A1
Havelock St 26 B2
Hawks La 26 A2
Hawthorn Av 26 B1
Heaton Rd 26 A3
High St 26 A2
High St 26 B2
Hollow La 26 A3
Hollowmede 26 A3
Honeywood Clo 26 B1
Hospital La 26 A2
Hudson Rd 26 B1
Iron Bar La 26 B2
Ivy La 26 B2
Ivy Ter 26 A3
Jackson Rd 26 A2
Jessica Mews 26 B1
Jewry La 26 A2
Juniper Clo 26 B3
Keyworth Mews 26 B1
King St 26 B2
Kingsmead Rd 26 B1
Kirbys La 26 A2
Knotts La 26 B2
Knowlton Walk 26 B1
Lady Woottons Gdns 26 B2
Lancaster Rd 26 A3
Langton La 26 B3
Lansdown Rd 26 B3
Lesley Av 26 B3
Leycroft Clo 26 A1
Lime Kiln Rd 26 A3
Lincoln Av 26 B3
Linden Gro 26 A2
Link La 26 B2
London Rd 26 A2
Long Acre Clo 26 A1
Long Market 26 B2
Longport 26 B2
Love La 26 B2
Lower Bridge St 26 B2
Lower Chantry La 26 B2
Lyndhurst Clo 26 A1
Malthouse Rd 26 B1
Mandeville Rd 26 A1
Market Way 26 B1
Marlowe Av 26 A2
Martindale Clo 26 B3
Martyrs Field Rd 26 A3
Mary Green Walk 26 B1
Maynard Rd 26 A3
Mead Way 26 A2
Mercery La 26 B2
Metcalf Mews 26 B1
Military Rd 26 B2
Mill La 26 B2
Milton Rd 26 B3
Monastery St 26 B2
Nackington Rd 26 B3
New Dover Rd 26 B2
New Ruttington La 26 B1
New St 26 A2
New St 26 A3
New Town St 26 B1
Norfolk Rd 26 A3
Norman Rd 26 B3
North Holmes Rd 26 B2
North La 26 A2
Northgate 26 B2
Notley St 26 B1
Nunnery Fields 26 B3
Nunnery Rd 26 B3
Nursery Walk 26 A1
Oaten Hill 26 B2
Oaten Hill Pl 26 B2
Old Dover Rd 26 B2
Old Ruttington La 26 B2
Orange St 26 B2
Orchard St 26 A2
Oxford Rd 26 A3
Palace St 26 B2
Parham Rd 26 B1
Payton Mews 26 B1
Petchell Mews 26 B1
Pilgrims Way 26 B2
Pin Hill 26 A2
Pine Tree Av 26 A1
Pound La 26 A2
Pretoria Rd 26 B2
Princes Way 26 A2
Priory of St Jacob 26 A3
Prospect Pl 26 B3
Providence Row 26 A3
Puckle La 26 B3
Queens Av 26 A2
Ramsey Clo 26 A1
Randolph Clo 26 B3
Raymond Av 26 B3
Redwood Clo 26 A1
Remsdon Mews 26 B1
Rheims Way 26 A2
Rhodaus Clo 26 A3
Rhodaus Town 26 A2
Ringwood Clo 26 A1
Riverdale Rd 26 B1
Rochester Av 26 B3
Roper Clo 26 A1
Roper Rd 26 A2
Rose La 26 B2
Roseacre Clo 26 A1
Roselands Gdns 26 A1
Rosemary La 26 A2
Rushmead Clo 26 A1
Ryde St 26 A1
St Alphege La 26 B2
St Augustines Rd 26 B3
St Dunstans Clo 26 A1
St Dunstans St 26 A1
St Dunstans Ter 26 A2
St Edmunds Rd 26 A2
St Georges La 26 B2
St Georges Pl 26 B2
St Georges St 26 B2
St Georges Ter 26 B2
St Gregorys Rd 26 B2
St Jacobs Pl 26 A3
St Johns La 26 A2
St Johns Pl 26 B1
St Lawrence 26 B3
St Lawrence Forstal 26 B3
St Lawrence Rd 26 B3
St Margarets St 26 A2
St Martins Av 26 B2
St Martins Pl 26 B2
St Martins Rd 26 B2
St Martins Ter 26 B2
St Marys St 26 A2
St Michaels Pl 26 A1
St Michaels Rd 26 A1
St Mildreds Pl 26 A3
St Peters Gro 26 A2
St Peters La 26 A2
St Peters Pl 26 A2
St Peters St 26 A2
St Radigunds Pl 26 B1
St Radigunds St 26 A2
St Stephens Clo 26 B1
St Stephens Ct 26 A1
St Stephens Footpath 26 A1
St Stephens Grn 26 A1
St Stephens Hill 26 A1
St Stephens Rd 26 A1
Salisbury Rd 26 A1
Seymour Pl 26 A3
Shaftesbury Rd 26 A1
Simmonds Rd 26 A3
Somner Clo 26 A1
South Canterbury Rd 26 B3
Spring La 26 B2
Stanmore Ct 26 B3
Starle Clo 26 B1
Station Road E 26 A2
Station Road W 26 A2
Stephenson Rd 26 A1
Stour St 26 A2
Stour Vw 26 B1
Stuart Ct 26 B3
Stuppington La 26 A3
Sturry Rd 26 B1
Sun St 26 B2
Teddington Clo 26 B1
Temple Rd 26 A1
The Borough 26 B2
The Causeway 26 A1
The Drive 26 B3
The Friars 26 A2
The Gap 26 B3
The Hoystings Clo 26 B3
The Paddock 26 B2
The Riddings 26 B2
Tower Way 26 A2
Tudor Rd 26 A3
Union Pl 26 B1
Union St 26 B1
Upper Bridge St 26 B2
Upper Chantry La 26 B2
Valley Rd 26 A3
Vernon Pl 26 B2
Verwood Clo 26 A1
Victoria Rd 26 A3
Victoria Row 26 B2
Wacher Clo 26 A1
Watling St 26 A2
Wells Av 26 B3
West Pl 26 A1
Westgate Court Av 26 A1
Westgate Gro 26 A2
Westgate Hill Rd 26 A2
White Horse La 26 A2
Whitehall Bridge Rd 26 A2
Whitehall Clo 26 A2
Whitehall Gdns 26 A2
Whitehall Rd 26 A2
Whitstable Rd 26 A1
Willow Clo 26 B1
Wincheap 26 A3
Wincheap Ind Est 26 A2
Winchester Gdns 26 B3
Woodville Clo 26 A3
York Rd 26 A3
Zealand Rd 26 A3

CAPEL LE FERNE

Albany Rd 18 A4
Albert Rd 18 A4
Alexandra Rd 18 A4
Avondale Rd 18 A4
Beatrice Rd 18 A4
Capel St 18 A4
Cauldham Clo 18 A4
Cauldham La 18 A3
Clarence Rd 18 A4
Elizabeth Dri 18 A3
Green La 18 A3
Helena Rd 18 B3
Lancaster Av 18 A3
New Dover Rd 18 A4
Old Dover Rd 18 A4
Seaview Clo 18 A4
Victoria Rd 18 A4
Winehouse La 18 B3

CHARING

Ashford Rd 23 B4
Burleigh Rd 23 A4
Canterbury Rd 23 A3
Church La 23 B4
Clearmont Dri 23 B3
Downs Clo 23 B3
Downs Way 23 A3
Haffenden Meadow 23 A3
Hither Field 23 A4
Hook La 23 A4
Maidstone Rd 23 A3
Pett La 23 B3
Pilgrims Ct 23 A4
Pilgrims Way 23 B3
Pluckley Rd 23 A4
Pym Ho 23 B4
Sayer Rd 23 A3
School Rd 23 A3
Station Rd 23 A4
The Glebe 23 A3
The High St 23 B4
The Hill 23 B3
The Moat 23 B4
Toll La 23 B4
Westwell La 23 B4
Wheeler Rd 23 A3
Woodbrook 23 B4

CHATHAM

Acre Clo 27 A3
Admiralty Ter 27 A1
Afghan Rd 27 A2
Albany Rd 27 B3
Albany Ter 27 A2
Albert Rd 27 A3
Alexandra Rd 27 B3
Alfred Clo 27 B3
Amherst Hill 27 A1
Ansell Av 27 B3
Arden St 27 B1
Athelstan Rd 27 A3
Balfour Rd 27 A3
Bank St 27 B2
Barrier Rd 27 A2
Batchelor St 27 A2
Beaconsfield Rd 27 A3
Belmont Rd 27 B2
Beresford Av 27 A3
Best St 27 A2
Bingley Rd 27 A2
Blenheim Av 27 A3
Block Ct 27 B3
Booth Rd 27 A3
Boundary Rd 27 A3
Brenchley Clo 27 A3
Bright Rd 27 B3
Brisbane Rd 27 B3
Britton St 27 B2
Brompton Clo 27 A1
Brompton Hill 27 A1

Holly Gdns	29 B2
Holly La	29 B2
Invicta Rd	29 A3
Irvine Dri	29 B3
Jennifer Gdns	29 B3
Kent Rd	29 A3
Laleham Gdns	29 A2
Laleham Rd	29 A3
Laureate Clo	29 A2
Leicester Av	29 B2
Lewis Cres	29 A1
Lister Rd	29 A3
Lonsdale Av	29 A2
Lyndhurst Av	29 A2
Lyngate Clo	29 B2
Madeira Rd	29 A2
Magnolia Av	29 B2
Marlowe Rd	29 B3
Millmead Av	29 B3
Millmead Gdns	29 B3
Millmead Rd	29 A3
Newgate Lwr Prom	29 A1
Newgate Prom	29 A1
Norfolk Rd	29 A2
Northdown Av	29 A2
Northdown Hill	29 B3
Northdown Park Rd	29 A2
Northdown Rd	29 A2
Northdown Way	29 B3
Northumberland Av	29 B2
Offley Clo	29 B2
Olave Rd	29 A3
Old Green Rd	29 B2
Omer Av	29 B2
Palm Bay Av	29 B1
Palm Bay Gdns	29 B1
Palmer Cres	29 B3
Park Cres Rd	29 A2
Park La	29 A2
Park Rd	29 A2
Percy Av	29 A1
Poets Corner	29 A2
Prices Av	29 A2
Princes Gdns	29 B2
Princes Walk	29 A2
Princess Margaret Av	29 B2
Queen Elizabeth Av	29 B3
Queens Prom	29 A1
Richmond Av	29 A2
Rosedale Rd	29 A2
Rutland Av	29 A2
Rutland Gdns	29 B2
St Anthonys Way	29 B2
St Christopher Clo	29 B3
St Dunstans Rd	29 A2
St Francis Clo	29 B3
St Marys Av	29 B3
St Michaels Av	29 B3
St Mildreds Rd	29 A2
St Pauls Rd	29 A2
St Peters Footpath	29 A3
St Peters Rd	29 A3
Saltwood Gdns	29 B2
Sarah Gdns	29 B3
Second Av	29 A1
Selbourne Rd	29 A3
Simon Av	29 B2
Stanley Rd	29 A2
Surrey Rd	29 A2
Sweyn Rd	29 A1
Swinford Gdns	29 B3
Taddy Gdns	29 B3
Talbot Rd	29 A2
Thanet Rd	29 A2
The Ridgeway	29 A2
Third Av	29 A1
Tomlin Dri	29 B3
Upper Dane Rd	29 A2
Victor Av	29 B2
Victoria Av	29 A2
Warwick Rd	29 A2
Wellesley Rd	29 A2
West Park Av	29 B2
Western Rd	29 B3
Wharfdale Rd	29 A2
Whitfield Av	29 B3
Wilderness Hill	29 A2
William Av	29 B3
Willow Clo	29 B2
Windsor Av	29 A2
Wyndham Av	29 A2

COXHEATH

Adams Clo	30 B2
Amherst Villas	30 A1
Amsbury Rd	30 A2
Burston Rd	30 B2
Capel Clo	30 B2
Chestnut Dri	30 B2
Clinton Clo	30 B2
Cobtree Rd	30 B2
Crittenden Bungalows	30 A1
Culpepper Rd	30 B2
Dane St	30 B2
Dean St	30 A1
Foremans Barn Rd	30 A2
Forstal La	30 B1
Gallants La	30 A1
Georgian Dri	30 B2
Gresham Rd	30 B2
Hanover Rd	30 B2
Heath Rd	30 A1
Heathside Av	30 B1
Huntington Rd	30 B2
Linden Rd	30 B1
Linton Gore	30 B2
Mill La	30 B1
North Cres	30 B1
North Folly Rd	30 A2
Orchard Clo	30 B1
Pembroke Rd	30 B2
Pippin Clo	30 B2
Russett Clo	30 B2
South Cres	30 B1
Springett Way	30 B1
Stockett La	30 B1
The Beacons	30 B2
The Valley	30 B2
Upper Hunton Hill	30 A2
Waverley Clo	30 B2
Westerhill Rd	30 B2
Westway Park	30 B2
Whitebeam Dri	30 B2
Wilberforce Rd	30 B2
Wilsons La	30 A1
Woodlands	30 B2
Workhouse La	30 B1

CRANBROOK

Angley Rd	31 A2
Angley Walk	31 B1
Bakers Cross	31 B2
Bank St	31 B2
Bramley Dri	31 B3
Brickenden Dri	31 B2
Brickenden Rd	31 B2
Brookside	31 B2
Carriers Rd	31 B2
Causton Rd	31 A2
Crane La	31 B2
Dobells	31 B2
Dorothy Av	31 B3
Frythe Clo	31 B2
Frythe Cres	31 B3
Frythe Walk	31 B3
Frythe Way	31 B3
Goddards Clo	31 A2
Golford Rd	31 B2
Goudhurst Rd	31 B1
Greenway	31 A3
Hawkhurst Rd	31 A3
Hendley Dri	31 A2
High St	31 A2
Hopes Rd	31 B2
Huntingdon Clo	31 B2
Jockey La	31 B2
Kirby Clo	31 B3
New Rd	31 A2
Oatfield Clo	31 A2
Oatfield Dri	31 A2
Orchard Way	31 A3
Pennyfields	31 B3
Quakers Dri	31 B1
Quakers La	31 B1
Rope Walk	31 A2
St Dunstans Walk	31 A2
Sheafe Dri	31 A2
Stone St	31 B2
Swifts Vw	31 B1
The Crest	31 B2
The Hill	31 B2
The Tanyard	31 B2
Tilsden La	31 B3
Tippens Clo	31 B2
Tower Meadow	31 B2
Turnden	31 A3
Turner Av	31 B3
Waterloo Rd	31 B2
Wheatfield Clo	31 A2
Wheatfield Dri	31 A2
Wheatfield Lea	31 A2
Wheatfield Way	31 A2
Willesley Gdns	31 B1
Windmill Cottages	31 B2

CUXTON

Bush Rd	32 A1
Charles Dri	32 A1
Harold Rd	32 A1
Hayley Clo	32 A1
Hillcrest Dri	32 B1
James Rd	32 A1
Ladywood Rd	32 A1
Nine Acres Rd	32 A1
Pilgrims Way	32 B1
Poplicans Rd	32 A1
Reginald Av	32 A1
Rochester Rd	32 A2
Stanford Way	32 B1
Station Rd	32 B1
Sundridge Hill	32 B1
The Glebe	32 B2
White Leaves Rise	32 A1
William Rd	32 B1
Wood St	32 A2
Woodhurst Clo	32 A2
Wouldham Rd	32 B2

DARENTH/SUTTON-AT-HONE

Arnolds La	89 A2
Axtane Clo	89 B3
Balmoral Rd	89 A2
Barfield	89 A3
Barton Rd	89 A3
Bennett Way	89 B1
Burnt House La	89 A1
Cedar Dri	89 A3
Church Rd	89 A2
Clement St	89 A2
Coombefield	89 B1
Coopers Clo	89 B3
Cross Rd	89 A1
Dairy Clo	89 A2
Darent Mead	89 A3
Darenth Hill	89 B1
Darenth Road Sth	89 A1
Dartford By-Pass	89 A3
Dartford Rd	89 A3
Devon Ct	89 A3
Devon Rd	89 A3
East Hill	89 B3
Gorringe Av	89 B3
Green Street Green Rd	89 B1
Hawley Rd	89 A1
Holmesdale Hill	89 B3
Holmesdale Rd	89 B3
Hotham Clo	89 A2
Keith Av	89 A3
Longmarsh View	89 A3
Main Rd	89 A2
Mallys Pl	89 B3
Mill Rd	89 A1
Mill Stone Clo	89 A3
Montgomery Rd	89 B3
New Rd	89 B3
Paddock Clo	89 B3
Parsonage La	89 A1
Pembroke Clo	89 A3
Prince Charles Av	89 B3
Roman Villa Rd	89 B1
Royal Rd	89 A1
Russel Pl	89 A3
Ship La	89 A3
Shirehall Rd	89 A1
Shrubbery Rd	89 B3
Sinclair Way	89 B1
Smythe Rd	89 A3
Station Rd	89 A3
Tallents Clo	89 A2
The Grange	89 B3
Towers Wood	89 B3
Victoria Dri	89 B3
Watchgate	89 A1
Watermill Way	89 B3
Wood La	89 B1

DARTFORD

Acacia Rd	33 A3
Allen Clo	33 A1
Anne of Cleves Rd	33 A2
Ash Rd	33 A3
Austen Gdns	33 B1
Bath Rd	33 B2
Beech Rd	33 A3
Berkeley Cres	33 B3
Blackman Clo	33 A3
Blake Gdns	33 B1
Blenheim Clo	33 A2
Blenheim Rd	33 A2
Bondfield Walk	33 B1
Brent La	33 B2
Brone Gro	33 B1
Bullace La	33 B2
Burnham Cres	33 A1
Burnham Rd	33 A1
Cairns Clo	33 A1
Carsington Gdns	33 A3
Cedar Rd	33 A3
Central Rd	33 B1
Chatsworth Rd	33 A1
Chestnut Rd	33 A3
Christchurch Rd	33 A2
Church Hill	33 B3
Colney Rd	33 B2
Cranford Rd	33 B3
Crayside Ind Est	33 A1
Cross Rd	33 A2
Cumberland Dri	33 B2
Darenth La	33 B2
Darenth Rd	33 B3
Dartford Rd	33 A2
Dartford Trade Park	33 A3
Dene Rd	33 B2
Derwent Clo	33 A3
Devonshire Av	33 A2
East Hill	33 B2
East Hill Dri	33 B2
Egerton Clo	33 A3
Elm Clo	33 A3
Elm Rd	33 A3
Essex Rd	33 A2
Farthing Clo	33 B1
Firman Rd	33 B2
Francis Rd	33 A1
Fulwich Rd	33 B2
Gainsborough Av	33 A2
Gladstone Rd	33 B2
Gordon Rd	33 A2
Great Queen St	33 B2
Green Banks	33 B3
Grosvenor Cres	33 A1
Hall Rd	33 B1
Hallford Way	33 B3
Hawley Rd	33 B3
Hawthorn Rd	33 A3
Hazel Rd	33 A3
Heath La	33 A3
Heath St	33 A2
Henderson Dri	33 B1
Herald Walk	33 B2
High St	33 B2
Highfield Rd	33 A2
Highfield Rd Nth	33 A2
Highfield Rd Sth	33 A2
Hill Rd	33 B3
Hilltop Gdns	33 B1
Holly Rd	33 A3
Holmleigh Av	33 B2
Home Gdns	33 B2
Home Orchard	33 B2
Hythe St	33 B1
Ingram Rd	33 B3
Instone Rd	33 B2
Joyce Green La	33 B1
Joyce Green Walk	33 B1
Junction Rd	33 B2
Kent Rd	33 B2
Keyes Rd	33 B1
King Edward Av	33 A2
Kingsridge Gdns	33 A2
Kingswood Clo	33 A2
Knights Manor Way	33 B2
Laburnam Av	33 A3
Larch Rd	33 A3
Laurel Clo	33 A3
Lavinia Rd	33 B2
Lawford Gdns	33 A2
Lawrence Hill Gdns	33 A2
Lawrence Hill Rd	33 A2
Lawson Rd	33 A1
Linden Av	33 A3
Links Vw	33 A3
Little Queen St	33 B2
Lowfield St	33 B3
Mallard Clo	33 B3
Maple Rd	33 A1
Marcet Rd	33 B2
Market Pl	33 B2
Market St	33 B2
Marlborough Rd	33 B2
Marsh St	33 B1
Mayfair Rd	33 A1
Mead Cres	33 B3
Mead Rd	33 B3
Meadowside	33 B3
Mill Pond Rd	33 B2
Miskin Rd	33 A3
Monks Orchard	33 A3
Morland Av	33 A1
Mount Pleasant Rd	33 A3
Myrtle Rd	33 A3
Nelson Rd	33 A2
Norman Rd	33 B3
North St	33 B2
Oakfield La	33 A3
Oakfield Park Rd	33 A3
Oakfield Pl	33 A3

Olive Rd	33 A3	Claremont Rd	34 A3

Olive Rd 33 A3
Orchard St 33 B2
Osterberg Rd 33 B1
Overy St 33 B2
Penney Clo 33 A2
Phoenix Pl 33 B2
Powder Mill La 33 B3
Princes Rd 33 A3
Priory Clo 33 A1
Priory Gdns 33 A2
Priory Hill 33 A2
Priory Pl 33 A2
Priory Rd 33 A1
Prospect Pl 33 B2
Raeburn Av 33 A1
Rayford Clo 33 A2
Riverside Way 33 B1
Roseberry Gdns 33 A2
Rosedene Ct 33 A2
Rowan Cres 33 A3
Rutland Clo 33 A2
St Albans Rd 33 B2
St James Pl 33 B2
St Martins Rd 33 B2
Sanctuary Clo 33 A2
Sandpit Rd 33 A1
Savoy Rd 33 A1
Sharp Way 33 B1
Shepherds La 33 A2
Shirley Clo 33 A1
Somerset Rd 33 A2
Somerville Rd 33 B2
Spielman Rd 33 B1
Spital St 33 B2
Spring Vale 33 A2
Stanham Rd 33 A1
Sterndale Rd 33 B2
Strickland Av 33 B1
Suffolk Rd 33 B2
Summerhill Rd 33 A2
Sycamore Rd 33 A3
Temple Hill 33 B2
Temple Hill Sq 33 B1
The Homestead 33 A2
The Spires 33 A3
Tower Rd 33 A2
Trafalgar Rd 33 B3
Trevelyan Clo 33 B1
Trevithick Dri 33 B1
Tufnail Rd 33 B2
Tyler Gro 33 B1
Vauxhall Pl 33 B2
Victoria Rd 33 B1
Waid Clo 33 B2
Walnut Tree Av 33 B3
Wellcombe Av 33 B1
Wellington Rd 33 A2
West Hill 33 A2
West Hill Dri 33 A2
West Hill Rise 33 A2
West Lodge Av 33 A2
West View Rd 33 B2
Westgate Rd 33 A2
Willow Rd 33 A3
Wilmot Rd 33 A1
Windermere Clo 33 A3
Wyvern Clo 33 A2
York Rd 33 B2

DEAL

Albert Av 34 A3
Albert Rd 34 A3
Albion Rd 34 B1
Alfred Row 34 B3
Alfred Sq 34 B2
Allenby Av 34 A3
Anchor La 34 A2
Ark La 34 A2
Athelstan Pl 34 A1
Beach St 34 B2
Beaconsfield Rd 34 A3
Beechwood Av 34 A3
Blenheim Rd 34 A3
Bowling Green La 34 A3
Brewer St 34 B2
Bridge Rd 34 B2
Bridgeside 34 A2
Britannia Rd 34 B1
Broad St 34 B3
Buckthorn Clo 34 A1
Bulwark Rd 34 B2
Cannon St 34 A2
Canute Rd 34 A1
Capstan Row 34 B2
Century Walk 34 A2
Chapel St 34 B2
Church Path 34 A3
Clanwilliam Rd 34 B3

Claremont Rd 34 A3
Clarence Pl 34 B2
College Rd 34 A2
Coppin St 34 B2
Cowper Rd 34 A3
Deal Castle Rd 34 B3
Deal Rd 34 A3
Dibden Rd 34 B2
Dida Av 34 A3
Dolphin Sq 34 B2
Douglas Ter 34 A3
Duke St 34 A2
Enfield Rd 34 B2
Ethelbert Rd 34 A1
Exchange St 34 B2
Farrier St 34 B2
Garden Walk 34 A2
George All 34 B2
George St 34 B2
Georges Pass 34 B2
Gilford Rd 34 A3
Gladstone Rd 34 A3
Godwyn Rd 34 A1
Golden St 34 B2
Golf Rd 34 A1
Golf Road Pl 34 A1
Grange Rd 34 A3
Granville St 34 A3
Griffin St 34 B2
Harold Rd 34 A1
Hengist Rd 34 A1
High St 34 B2
Hope Rd 34 A3
Horsa Rd 34 B1
Ivy Pl 34 B2
Jernon Pl 34 B1
King Edward Av 34 A1
King St 34 B2
Leas Rd 34 A3
Links Rd 34 A1
Lister Clo 34 A3
London Rd 34 A3
Marine Rd 34 B3
Market St 34 B2
Matthews Clo 34 A2
Middle St 34 B2
Mill Rd 34 A3
Nelson St 34 A2
New St 34 B2
North Lea 34 A2
North St 34 B2
Northcote Rd 34 B3
Northwall Rd 34 A2
Oak St 34 B2
Park Av 34 A3
Park St 34 A3
Peter St 34 A2
Portobello Ct 34 B2
Prince of Wales Ter 34 B3
Princess St 34 A2
Queen St 34 A3
Ranelagh Rd 34 B3
Ravenscourt Rd 34 A3
Redhouse Wall 34 A1
Roberts St 34 A2
St Andrews Rd 34 A2
St Davids Rd 34 A2
St Georges Rd 34 A2
St Leonards Rd 34 A3
St Patricks Clo 34 A2
St Patricks Rd 34 A2
Sandown Rd 34 B1
Saxon Pl 34 A1
Sholden Church La 34 A3
Silver St 34 B2
Sondes Rd 34 B3
South Par 34 B2
South St 34 B3
South Wall 34 A2
Southwall Rd 34 A2
Stanhope Rd 34 A2
Stanley Rd 34 B3
Sutherland Rd 34 A3
Sydenham Rd 34 B2
Tar Path 34 A3
The Drive 34 A3
The Fairway 34 A1
The Grove 34 A3
The Marina 34 B1
The Strand 34 B3
Union Rd 34 A2
Victoria Mews 34 A3
Victoria Par 34 B3
Victoria Rd 34 B3
Water St 34 B2
Wellington Rd 34 A3
West Lea 34 A2
West St 34 A2
Western Rd 34 A2
William Pitt Av 34 A3
Wilton Rd 34 A3

DOVER

Adrian St 35 A2
Albany Pl 35 A2
Albert Rd 35 A1
Archcliffe Rd 35 A3
Ashen Tree La 35 A1
Athol Ter 35 B1
Avenue Rd 35 A1
Bartholomew St 35 A1
Barton Path 35 A1
Bastion Rd 35 A2
Beaconsfield Av 35 A1
Beaconsfield Rd 35 A1
Bench St 35 A2
Biggin St 35 A1
Bowling Green Ter 35 A2
Branch St 35 A1
Bridge St 35 A1
Bulwark St 35 A3
Cambridge Rd 35 A2
Camden Cres 35 A2
Cannon St 35 A2
Canons Gate Rd 35 B1
Castle Av 35 A1
Castle Hill Rd 35 B2
Castle St 35 A2
Castlemount Rd 35 A1
Centre Rd 35 A3
Channel View Rd 35 A3
Chapel La 35 A2
Chapel Pl 35 A2
Charlton Centre 35 A1
Charlton Grn 35 A1
Church St 35 A2
Churchill St 35 A1
Citadel Rd 35 A3
Clarence Pl 35 A3
Clarendon Rd 35 A2
Connaught Rd 35 A1
Constables Rd 35 B1
Cowgate Hill 35 A2
Crafford St 35 A1
De Burgh St 35 A1
Deal Rd 35 B1
Dieu Stone La 35 A2
Dolphin La 35 A2
Dolphin Pl 35 A2
Dour St 35 A1
Douro Pl 35 B2
Drop Redoubt Rd 35 A2
Durham Clo 35 A2
Durham Hill 35 A2
East Cliff 35 B2
East Norman Rd 35 B1
East Roman Ditch 35 B1
Edwards Rd 35 A1
Effingham Cres 35 A1
Effingham Pass 35 A2
Effingham St 35 A2
Elizabeth St 35 A3
Esplanade 35 A3
Fishmongers La 35 A2
Flyinghorse La 35 A2
Folkestone Rd 35 A2
Fort Burgoyne Rd 35 B1
Frith Rd 35 A1
Gaol La 35 A2
Godwin 35 B1
Godwyne Clo 35 A1
Godwyne Path 35 A1
Godwyne Rd 35 A1
Goodfellow Way 35 A1
Granville St 35 A1
Harold Pass 35 B1
Harold St 35 A1
Harolds Rd 35 B1
Hawkesbury St 35 A3
Hewitt Rd 35 A1
High St 35 A1
Jubilee Way 35 B2
King St 35 A2
Knights Rd 35 B1
Knights Templars 35 A3
Ladywell 35 A1
Lancaster Rd 35 A2
Laureston Pl 35 B1
Leyburne Rd 35 A1
Limekiln St 35 A3
London Rd 35 A1
Lord Warden Sq 35 A3
Maison Dieu Pl 35 A1
Maison Dieu Rd 35 A1
Malvern Rd 35 A2
Marine Par 35 B2
Market Sq 35 A2
Matthews Pl 35 A1
Military Rd 35 A2
Mill La 35 A2
Mortimer Rd 35 B1
New Bridge 35 A2

New St 35 A2
Norman St 35 A2
North Military Rd 35 A2
Park Av 35 A1
Park Mews 35 A1
Park Pl 35 A1
Park St 35 A1
Pencester Rd 35 A2
Peter St 35 A1
Princes St 35 A2
Priory Gate Rd 35 A1
Priory Gro 35 A1
Priory Hill 35 A1
Priory Pl 35 A2
Priory Rd 35 A1
Priory Station App Rd 35 A2
Priory St 35 A2
Promenade 35 B2
Queen Elizabeth Rd 35 B2
Queen St 35 A2
Queens Gdns 35 A2
Russell St 35 A2
St Alphege Rd 35 A1
St James St 35 A2
St Johns Rd 35 A2
St Martins Hill 35 A3
St Martins Path 35 A3
St Martins Steps 35 A2
St Marys Passage 35 A2
St Pauls Pl 35 A1
Salisbury Rd 35 A1
Saxon St 35 A2
Snargate St 35 A3
South Military Rd 35 A3
Stenbrook 35 A2
Strond St 35 A3
Taswell St 35 A1
The Gateway 35 B2
The Paddock 35 A1
The Spur 35 A3
The Viaduct 35 A3
Tower Hamlets Rd 35 A1
Townwall St 35 A2
Union St 35 A3
Upper Rd 35 B1
Victoria Cres 35 A1
Victoria Pk 35 B1
Waterloo Cres 35 A2
Wellesley Rd 35 A2
West Norman Rd 35 B1
West Roman Ditch 35 B1
West Wing Rd 35 B1
Wood St 35 A1
Woolcomber St 35 B2
Worthington St 35 B2
York St 35 A2

DUNTON GREEN/RIVERHEAD

Amherst Hill 36 B2
Anthony Clo 36 B1
Baden Powell Rd 36 B2
Barnfield Rd 36 A2
Barretts Rd 36 B1
Bessels Green Rd 36 A3
Bessels Meadow 36 A3
Bessels Way 36 A3
Betenson Av 36 B2
Bradbourne Vale Rd 36 B2
Braeside Av 36 B3
Braeside Clo 36 B3
Brittains La 36 B3
Broomfield Rd 36 B2
Bullfinch Clo 36 A2
Bullfinch Dene 36 A2
Bullfinch La 36 A2
Chesterfield Dri 36 A2
Chipstead La 36 A2
Chipstead Park 36 A2
Chipstead Park Clo 36 A2
Chipstead Place Gdns 36 A2
Church Field 36 B2
Cold Arbour Rd 36 B3
Court Rd 36 B3
Court Wood Dri 36 B2
Cranmer Rd 36 B2
Crawshaw Clo 36 B2
Crescent Cottages 36 B1
Croftway 36 B3
Darenth Clo 36 A2
Darenth La 36 B1
Donnington Rd 36 A1
Downsview Rd 36 B3
Elmstead Clo 36 B2
Greenwood Way 36 B3
Hamlin Rd 36 B2
Hawthorn La 36 B2
Heatherfield Rd 36 B2
High St 36 A2

Hillfield Rd	36 B1
Homedean Rd	36 A2
Homefield Rd	36 B2
Kingswood Rd	36 B1
Kippington Rd	36 B3
Lake View Rd	36 B2
Larkfield Rd	36 B3
Lennard Rd	36 B1
Linden Sq	36 B2
London Rd	36 B1
Lusted Rd	36 B1
Lyndhurst Dri	36 B3
Madison Way	36 B2
Maidstone Rd	36 B2
Marlborough Cres	36 B3
Martins Shaw	36 A2
Middlings Wood	36 B3
Mill Rd	36 B1
Milton Rd	36 B1
Montreal Rd	36 B2
Morewood Clo	36 B2
Mount Clo	36 B2
Nursery Pl	36 A2
Old Carriageway	36 A2
Orchard Rd	36 B2
Packhorse Rd	36 A2
Park Pl	36 A3
Pontoise Clo	36 B2
Poundsley Rd	36 B1
Quarry Cotts	36 B3
Redlands Rd	36 B3
River Par	36 B2
Robyns Way	36 B2
Rose Field	36 B3
Rye La	36 B1
St Marys Dri	36 B2
Sandilands	36 A2
Scott Way	36 B2
Sevenoaks By-Pass	36 A3
Shoreham La	36 B2
Springshaw Clo	36 A2
Stairfoot La	36 A2
Stanhope Way	36 A2
Stapleford Ct	36 B3
Station Rd	36 B1
Sunrise Cotts	36 B1
The Close	36 B3
The Meadway	36 B2
The Old Garden	36 A2
The Patch	36 B2
The Square	36 B2
The Terrace	36 A2
Uplands Clo	36 B2
Uplands Way	36 B2
Vicarage La	36 A1
Westerham Rd	36 A2
Westwood Way	36 B2
Whitehart Par	36 B2
Witches La	36 A2
Woodfields	36 A2
Worships Hill	36 A2
Yew Tree Clo	36 A2

DYMCHURCH

Burmarsh Rd	37 B1
Chapel Rd	37 B2
Crossways Clo	37 C1
Dunkirk Clo	37 B2
Eastbridge Rd	37 A2
Green Meadows	37 C1
High St	37 B2
Hind Clo	37 B2
Hythe Rd	37 B1
Kingsway	37 C1
Lower Sands	37 C1
Lyndhurst Rd	37 B2
Marshlands	37 B2
Marshlands Clo	37 B2
Mill Rd	37 B2
Mitcham Rd	37 B2
New Hall Clo	37 B2
Orgarswick Av	37 B2
Orgarswick Way	37 B2
Queensway	37 C1
Rush Clo	37 B2
St Anns Rd	37 B2
St Marys Rd	37 A2
Sark Clo	37 B2
Sea Wall	37 B2
Seabourne Way	37 B2
Ship Clo	37 B2
Station Rd	37 B2
Sycamore Clo	37 B2
Sycamore Gdns	37 B2
Tartane La	37 B2
The Oval	37 C1
Tower Est	37 C1
Tritton Gdns	37 C1
Tudor Av	37 B1
Venture Clo	37 C1

Wraights Field	37 B2

EAST PECKHAM

Addlestead Rd	38 A2
Bardsley Clo	38 B1
Barnfield	38 B1
Bramley Rd	38 A1
Branbridges Rd	38 B2
Bullen La	38 A1
Bush Rd	38 A1
Chidley Cross Rd	38 A1
Church La	38 A1
Cotman Way	38 A1
Crown Acres	38 B2
Drage Rd	38 A1
Fell Mead	38 A2
Freehold	38 B2
Golding Gdns	38 B1
Hale Ct	38 B1
Hale St	38 B1
Hatches La	38 A1
Henham Gdns	38 B1
Marvillion Ct	38 B1
Medway Meadows	38 B1
Old Rd	38 B2
Orchard Rd	38 B2
Peckham Ct	38 B2
Pinkham Gdns	38 B2
Pippin Rd	38 A1
Pound Rd	38 A1
Russet Rd	38 A1
Smithers Ct	38 B1
Smithers La	38 B1
Snoll Hatch La	38 A2
Stockenbury	38 A2
Strettitt Gdns	38 B2
Tonbridge Rd	38 A2
Torbay Rd	38 B2
Twysden Ct	38 B1
Westwood Rd	38 A2
Whitebine Gdns	38 B1
William Luck Clo	38 A1

EASTRY

Albion Rd	38 A4
Boystown Pl	38 B3
Brook St	38 B4
Church St	38 B4
Cooks Lea	38 A4
Gore La	38 A4
Gore Rd	38 A3
Hay Hill	38 B4
High St	38 B4
Liss Rd	38 A4
Lower Gore La	38 A3
Lower St	38 B4
Mill Grn	38 A4
Mill La	38 A4
Orchard Rd	38 A3
Peak Dri	38 A3
Sandwich Rd	38 B3
Selson La	38 A3
Statenborough La	38 B3
Swaynes Way	38 A4
Wheelwrights Way	38 A4
Woodnesborough La	38 B3

EDENBRIDGE

Albion Way	39 A1
Ashcombe Dri	39 A1
Barn Hawe	39 A3
Briar Clo	39 A3
Cedar Dri	39 A2
Chestnut Clo	39 A2
Church St	39 B3
Churchfield	39 B3
Clover Walk	39 B2
Commerce Way	39 A2
Coomb Field	39 A3
Croft Ct	39 B3
Croft La	39 B3
Crouch House Rd	39 A2
Crown Rd	39 B1
Enterprise Way	39 A1
Faircroft Way	39 A2
Fairmead Rd	39 A1
Farmstead Dri	39 B2
Field Dri	39 B2
Forge Croft	39 B3
Four Elms Rd	39 A2
Foxglove Clo	39 B2
Frantfield	39 B3
Grange Clo	39 B2
Great Mead	39 A2
Greenfield	39 B3
Harrow Clo	39 B2
Hawthorn Clo	39 A2
Heron Clo	39 B2
Hever Rd	39 B3

High Fields Rd	39 A1
High St	39 A2
Hilders Clo	39 A1
Hilders La	39 A1
Hillcrest	39 A1
Homestead Rd	39 A1
Hopgarden Clo	39 B2
Katherine Rd	39 B3
Kestrel Clo	39 B2
Lingfield Rd	39 A3
Lucilina Dri	39 A3
Lynmead Clo	39 A1
Magpie Grn	39 B2
Main Rd	39 A1
Mallard Way	39 B2
Manor House Gdns	39 A3
Manor Rd	39 A3
Marl Hurst	39 A1
Marlpit Clo	39 A1
Meadow La	39 A1
New House Ter	39 A2
Oak Vw	39 A2
Oakfield Rd	39 A1
Orchard Clo	39 A2
Orchard Dri	39 A2
Oxfield	39 B2
Park Av	39 A2
Park View Clo	39 A2
Penlee Clo	39 A2
Pine Gro	39 A2
Pit La	39 A1
Plough Walk	39 B2
Plover Clo	39 B2
Queens Ct	39 B3
Ridge Way	39 B1
Riverside Ct	39 B3
Riverside	39 B3
Robyns Way	39 B2
Rowfield	39 B2
Skeynes Rd	39 A3
Skinners La	39 B2
Smithyfield	39 B2
Sorrell Clo	39 B2
Springfield Rd	39 A3
Stackfield	39 B2
Stanbridge Rd	39 A2
Stangrove Rd	39 A3
Station App	39 A2
Station Rd	39 A2
Stoneyfield	39 B2
Streatfield	39 B3
Swan La	39 A1
Swan Ridge	39 B1
The Limes	39 A3
The Plat	39 B3
Victoria Clo	39 B3
Victoria Rd	39 B3
Wainhouse Clo	39 B2
Water Lakes	39 B3
Wayside Dri	39 B2
Westways	39 A2
Woodland Dri	39 B2
Woodpecker Clo	39 B2

FAVERSHAM

Abbey Pl	40 B2
Abbey Rd	40 B2
Abbey St	40 B2
Aldred Rd	40 B3
Alexander Dri	40 A2
Arthur Salmon Clo	40 A2
Ashford Rd	40 B3
Athelstan Rd	40 A3
Bank St	40 B2
Barnes Clo	40 A2
Barnfield Rd	40 B2
Beckett St	40 B2
Beech Clo	40 A2
Belmont Rd	40 B3
Belvedere Rd	40 B2
Bensted Gro	40 A2
Blaxland Clo	40 A2
Bramblehill Rd	40 B2
Brent Hill	40 B2
Brent Rd	40 B2
Bridge Rd	40 B2
Briton Rd	40 B3
Brogdale Rd	40 A3
Brook Rd	40 B2
Broomfield Rd	40 B2
Cambridge Rd	40 A3
Canute Rd	40 B3
Capel Rd	40 A3
Caslocke	40 B2
Cavour Rd	40 B2
Chapel St	40 B3
Chart Clo	40 A2
Church Rd	40 B2
Church Rd, Oare	40 A1
Church St	40 B2

Churchill Way	40 A2
Cobb Walk	40 A2
Cobham Chase	40 A2
Colegates Clo	40 A1
Colegates Rd	40 A1
Conduit St	40 B2
Court St	40 B2
Crescent Rd	40 B2
Cress Way	40 A2
Crispin Clo	40 B2
Cross La	40 B2
Dark Hill	40 A2
Davington Hill	40 B2
Dorset Pl	40 B3
East St	40 B2
Edith Rd	40 B3
Egbert Rd	40 B3
Ethelbert Rd	40 A3
Everard Way	40 A2
Fielding St	40 B2
Finlay Clo	40 A2
Flood La	40 B2
Forbes Rd	40 B3
Forstall Rd	40 B2
Front Brents	40 B2
Garfield Pl	40 B2
Gatefield La	40 B2
Giraud Dri	40 A2
Goldfinch Clo	40 B1
Granville Clo	40 B2
Ham Rd	40 B2
Harold Rd	40 B3
Harrison Ter	40 A1
Hatch St	40 B2
Hazebrouk Rd	40 A2
Horsford Walk	40 A2
Hugh Pl	40 B2
Institute Rd	40 B2
John Hall Clo	40 A1
Johnson Ct	40 A1
Judd Rd	40 A2
Kennedy Clo	40 A2
Kiln Ct	40 A2
Kings Rd	40 B2
Kingsnorth Rd	40 B3
Larksfield Rd	40 A3
Lion Field	40 A3
London Rd	40 A3
Lower Rd	40 A2
Maitland Ct	40 A1
Makenade Av	40 B3
Market Pl	40 B2
Memdfield St	40 B2
Middle Row	40 B2
Millstream Clo	40 B2
Monks Dri	40 A2
Mount Field	40 A3
Mount Pleasant	40 A1
Mutton La	40 A3
Napleton Rd	40 B2
Nelson St	40 B3
Newton Rd	40 B3
Nightingale Rd	40 B3
Nobel Ct	40 A2
Norman Rd	40 B2
North La	40 B2
Oare Rd	40 A1
Old Gate Rd	40 B2
Orchard Pl	40 B2
Ospringe Pl	40 A3
Ospringe Rd	40 A3
Ospringe St	40 A3
Park Rd	40 B3
Partridge La	40 A2
Penshurst Ridge	40 A2
Plantation Rd	40 B3
Preston Gro	40 B3
Preston La	40 B3
Preston Park Est	40 B3
Preston St	40 B3
Priory Pl	40 B2
Priory Rd	40 B2
Priory Row	40 B2
Quay La	40 B2
Queens Rd	40 A3
Reedland Cres	40 B2
Roman Rd	40 B3
Russell Pl	40 A1
St Anns Rd	40 A3
St Catherines Dri	40 B3
St Johns Rd	40 B3
St Marys Rd	40 B3
St Nicholas Rd	40 A2
St Pauls Av	40 A2
St Peters Ct	40 A3
Salters La	40 B3
Saxon Rd	40 B3
School Rd	40 A3
Seager Rd	40 A1
Seven Acre	40 B2
Sherways	40 A3

Rectory Rd 44 C2
Riverview Rd 44 A1
Southfleet Rd 44 C1
Spring Vale 44 A1
Stanhope Rd 44 C1
Stanley Rd 44 C1
Starboard Av 44 A1
Station Rd 44 A1
Sun Rd 44 C1
Swanscombe St 44 C1
Sweyne Rd 44 A1
The Avenue 44 A1
The Crescent 44 A1
The Grove 44 C1
Trebble Rd 44 B1
Valley Vw 44 A1
Vernon Rd 44 C1
Wallace Gdns 44 B1
Watling St 44 A2
Wright Clo 44 B1

HADLOW

Appletons 45 B2
Bell Row 45 B3
Blackmans La 45 A2
Bourne Grange La 45 A2
Broadway 45 A2
Brookfields 45 A2
Carpenters La 45 A1
Caxton La 45 A2
Cemetery La 45 B2
Chesfield Clo 45 B2
Church St 45 A2
Common Rd 45 A1
Court La 45 B2
Great Elms 45 A1
Hailstone Clo 45 A2
Hartlake La 45 B3
High St 45 A2
Hope Av 45 A1
Kelchers La 45 B3
Kenward Rd 45 A2
Leeds House Mews 45 B2
Lonewood Way 45 B1
Maidstone Rd 45 B2
Maltings Clo 45 A2
Marshall Gdns 45 A1
Medway Vw 45 B3
Park Villas 45 B1
School La 45 A2
Sheraton Park 45 B3
Smithers Clo 45 B1
Steers Pl 45 A1
Tainter Rd 45 A1
The Cherry Orchard 45 A1
The Freehold 45 A1
The Maltings 45 A2
The Paddock 45 A1
Three Elm La 45 A3
Toby Gdns 45 A2
Tonbridge Rd 45 A2
Valley Dri 45 B1
Victoria Rd 45 B3
Water Slippe 45 A1
Wyford Rd 45 A1

HALLING

Ashby Clo 32 B4
Barn Meadow 32 A4
Bradley Rd 32 A4
Britannia Clo 32 B4
Browndens Rd 32 A4
Cemetery Rd 32 B4
Chapel La 32 A4
Chillington Clo 32 A4
Essex Rd 32 B3
Ferry Rd 32 B4
Formby Rd 32 B3
Grove Rd 32 A3
Halling By-Pass 32 B3
High St 32 B3
Kent Rd 32 B3
Lambarde Clo 32 B4
Marsh Rd 32 B3
Meadow Clo 32 A4
Meadow Cres 32 A4
Pilgrims Way 32 A3
Primrose Rd 32 A3
Stake La 32 B3
Station App 32 B3
The Street 32 A4
Vicarage Clo 32 B3
Vicarage Rd 32 A3

HAM STREET

Bourne La 102 A3
Bourne Wood 102 A3
Bunkley Meadow 102 A4

Carters Wood 102 A3
Cock La 102 A4
Dukes Meadow 102 A4
Fairfield Ter 102 A4
Farm Rd 102 A4
Hamstreet Rd 102 A4
Quince Orchard 102 A3
Romney Marsh Rd 102 A4
Romney Rd 102 A4
Ruckinge Rd 102 A3
St Marys Clo 102 A3
The Street 102 A3
Village Way 102 A4
Willow Dri 102 A4

HARRIETSHAM

Ashford Rd 46 B1
Church Cres 46 B1
Church La 46 B1
Church Rd 46 B1
Court Lodge La 46 A1
Cricketers Clo 46 A1
East St 46 A1
Fairbourne La 46 A2
Forge Meadow 46 A1
Harrietsham By-Pass 46 A1
Hook La 46 A1
Ivens Way 46 A1
Lakelands 46 B1
Marley Rd 46 B1
Mercer Dri 46 B1
North Downs Vw 46 B1
Old Layne 46 B1
Quested Way 46 A1
Rectory La 46 B2
St Welcumes Way 46 B1
Sandway Rd 46 B2
Station Rd 46 A1
Stede Hill 46 B1
West St 46 A1

HAWKHURST

All Saints Rd 46 B4
Barretts Rd 46 A3
Basden Gdns 46 B3
Copthall Av 46 A4
Cranbrook Rd 46 A3
Dunlop Ct 46 B3
Eden Ct 46 A3
Fairview 46 A4
Fieldways 46 B4
Hammonds 46 B3
Hartnokes 46 B3
Heartenoak Rd 46 B3
High St 46 A3
Highfield Clo 46 A4
Highgate Hill 46 A4
Mercers 46 A4
Murton Neale Clo 46 B3
Northgrove Rd 46 A3
Oakfield 46 A3
Oaklands Rd 46 A4
Ockley La 46 A3
Ockley Rd 46 A3
Park Cotts 46 B3
Queens Ct 46 B3
Queens Rd 46 B3
Rye Rd 46 B3
School Ter 46 A3
Slip Mill Rd 46 A3
Sopers Rd 46 A3
Tates 46 A4
The Colonnade 46 A3
The Smugglers 46 B4
Theobalds 46 A3
Vale Rd 46 A3
Water La 46 B4
Western Av 46 A3
Western Rd 46 A3
Whites La 46 B3
Winchester Rd 46 A3
Woodbury Rd 46 A3

HEADCORN

Ashleigh Gdns 47 A1
Bankfields 47 A1
Bramleys 47 B2
Brooklands 47 A1
Chaplin Dri 47 A2
Church Path 47 A2
Clerksfield 47 A2
Dawks Meadow 47 A2
Forge La 47 A2
Forge Meadows 47 A1
Gibbs Hill 47 B2
Gooseneck La 47 A2
Griggs La 47 B2

High St 47 A2
Kings Rd 47 A1
Kingsland Gro 47 B2
Knaves Acres 47 A1
Knights Way 47 A1
Knowles Gdns 47 B2
Lenham Rd 47 A1
Maidstone Rd 47 A1
Millbank 47 A1
Moat Rd 47 A1
New Rd 47 B2
North St 47 A1
Oak Farm Gdns 47 A1
Oak La 47 A1
Oakfields 47 A1
Orchard Glade 47 B2
Rushford Clo 47 A2
Station App 47 A2
Station Rd 47 A2
Thatch Barn Rd 47 A1
Tollgate Pl 47 B2
Ulcombe Rd 47 A1
Wheeler St 47 B2
Youngs Pl 47 A1

HERNE BAY

Albany Dri 47 A3
Arkley Rd 47 B4
Ashtrees 47 A3
Avenue Rd 47 A3
Bank St 47 B3
Beach St 47 A3
Beacon Hill 47 B3
Beacon Rd 47 B3
Beacon Walk 47 B3
Beaumanor 47 B4
Belle View Rd 47 B3
Beltinge Rd 47 B3
Bognor Dri 47 A4
Bowes La 47 B4
Brunswick Sq 47 A3
Bullers Av 47 A3
Bullockstone Rd 47 A4
Burton Fields 47 B4
Canterbury Rd 47 B4
Cavendish Rd 47 B3
Cecil Pk 47 B3
Cecil St 47 B3
Central Parade 47 A3
Chapel St 47 B3
Charles St 47 B3
Cherry Gdns 47 A4
Clarence Rd 47 A3
Clarence St 47 A3
Cobblers Bridge Rd 47 A4
Collard Clo 47 B4
Coopers Hill 47 B3
Courtlands 47 A4
Cross St 47 A4
Darenth Clo 47 B4
Dence Clo 47 B3
Dence Pk 47 B3
Dering Rd 47 A3
Dolphin St 47 A3
Douglas Rd 47 B4
Downs Pk 47 B3
East Gate Clo 47 B4
East St 47 B3
Eddington La 47 A4
Elizabeth Way 47 B4
Fernlea Av 47 A4
Fleetwood Av 47 A4
Gordon Rd 47 A4
Gosfield Rd 47 B3
Green Acres Clo 47 B4
Greenhill Rd 47 A4
Hadleigh Gdns 47 B3
Hanover Sq 47 A3
Hanover St 47 A3
Herne Av 47 B4
Herneville Gdns 47 B4
High St 47 B3
Hill Borough Rd 47 B3
Hillbrow Av 47 B4
Hillcroft Rd 47 B4
Hilltop Rd 47 B3
Ivanhoe Rd 47 B4
Kingfisher Clo 47 A4
Kings Rd 47 A3
Lane End 47 A3
Leighville Dri 47 A3
Linden Av 47 A4
Links Clo 47 B4
Little Charles St 47 B3
Margate Ct, Kings Rd 47 A3
Market St 47 A3
Mayfield Rd 47 B4
Mickleburgh Av 47 B4
Mickleburgh Hill 47 B3
Mill La 47 B4

Minster Dri 47 A3
Montague St 47 A3
Mortimer St 47 A3
New St 47 B3
North St 47 B3
Oakdale Rd 47 B4
Orchard Rd 47 B4
Oxend St 47 A3
Oxenden Park Dri 47 A4
Oxenden Sq 47 A4
Park Rd 47 A3
Parsonage Rd 47 B4
Pettman Clo 47 B4
Pier Av 47 A3
Pier Chine 47 A3
Pigeon La 47 B4
Priory La 47 B4
Prospect Hill 47 B3
Queen St 47 A3
Queens Gdns 47 B3
Ravensbourne Rd 47 B4
Reynolds Clo 47 B4
Richmond St 47 A3
Roselea Av 47 B4
St Andrews Clo 47 B4
St Annes Dri 47 A4
St Georges Ter 47 A3
Sandown Dri 47 A3
Sea St 47 A3
Seaview Sq 47 A3
South Rd 47 B3
Southsea Dri 47 A4
Spenser Rd 47 A4
Stanley Gdns 47 B4
Stanley Rd 47 B4
Station Chine 47 A4
Station Rd 47 B4
Swale Clo 47 B4
Telford St 47 A3
Thanet Way 47 A3
The Broadway 47 A3
The Circus 47 A4
The Downings 47 B4
Thunderland Rd 47 B4
Tower Gdns 47 A3
Tyndal Pk 47 B3
Underdown La 47 B4
Underdown Rd 47 B3
Victoria Pk 47 B3
Western Av 47 A3
Western Esplanade 47 A3
William St 47 B3
York Clo 47 A3
York Rd 47 A3

HIGHAM

Ash Cres 48 A1
Beech Gro 48 A1
Briar Dale 48 A1
Brice Rd 48 A2
Carton Rd 48 A1
Charles Dickens Av 48 A2
Chilton Dri 48 A2
Copperfield Cres 48 A2
Crutches La 48 A2
Dombey Clo 48 A2
Elm Clo 48 A2
Fairview Dri 48 A1
Forge La 48 A2
Gravesend Rd 48 A2
Hayes Clo 48 A2
Hermitage Rd 48 A1
High Vw 48 A1
Highwood Clo 48 A1
Hollytree Dri 48 A1
Irvine Rd 48 A2
Mountbatten Av 48 A1
Norah La 48 A1
Oak Dri 48 A1
Peartree La 48 A2
Peggoty Clo 48 A2
St Johns Clo 48 A1
St Johns Rd 48 A1
School La 48 A1
Taylors La 48 A1
Telegraph Hill 48 A2
The Braes 48 A2
The Larches 48 A1
Villa Rd 48 A2
Walmers Av 48 A1

HIGH HALSTOW

Britannia Rd 48 B3
Christmas La 48 B4
Churchill Pl 48 B3
Clinch St 48 B3
Cooling Rd 48 A3
Deangate Rd 48 A4

Dux Court Rd	48 A4
Eden Rd	48 B3
Forge La	48 B4
Goodwood Clo	48 B4
Harrison Dri	48 B3
Hill Farm Clo	48 B4
Hill View Cotts	48 B4
Longfield Av	48 B3
Marsh Cres	48 B3
Medway Av	48 B3
Northwood Av	48 B3
St Margarets Clo	48 B4
Thames Av	48 B3
Thames Vw	48 A4
The Street	48 B4
Willowbank Dri	48 B3

HILDENBOROUGH

Ashley Rd	49 B3
Birch Clo	49 B3
Bramble Clo	49 B3
Brookmead	49 B3
Byrneside	49 B3
Church Rd	49 B2
Coldharbour La	49 B2
Copse Rd	49 B3
Derby Clo	49 B3
Elm Gro	49 B3
Fairfield Way	49 B2
Fellowes Way	49 B2
Fir Tree Lodge	49 B2
Foxbush	49 A1
Francis Rd	49 B1
Garlands	49 B1
Greenview Cres	49 B3
Half Moon La	49 B2
Hardwick Rd	49 B1
Knowsley Way	49 B2
Leigh Rd	49 B3
Leybank	49 B3
London Rd	49 A1
Meadway	49 B2
Mill Lane	49 A1
Mount Pleasant	49 B2
Mount Pleasant Ct	49 B2
Noble Rd	49 A2
Orchard Lea	49 B2
Riding La	49 B2
Ridings Pk	49 B1
Rings Hill	49 A2
Stocks Green Rd	49 A2
Tonbridge By-Pass	49 A2
Tonbridge Rd	49 A2
Watts Cross Rd	49 A2
Wealdon Clo	49 B3
Wilson Clo	49 B2
Woodfield Av	49 B2
Woodview Cres	49 B3

HOO
ST. WERBURGH

Armytage Clo	50 B2
Aveling Clo	50 A1
Bells La	50 A1
Blackman Clo	50 A1
Brook Side	50 B2
Butt Haw Clo	50 B2
Church St	50 B2
Coombe Rd	50 B2
Everest Dri	50 B2
Everest Mews	50 B2
Forewents Rd	50 A1
Gordon Rd	50 A2
Herds Down	50 A2
Killick Rd	50 A2
Kingshill Dri	50 A1
Kingsnorth Clo	50 B1
Knights Clo	50 A1
Knights Rd	50 A2
Linton Dann Clo	50 A1
Main Rd	50 A2
Marley Rd	50 A1
Miskin Rd	50 A2
Morement Rd	50 A1
Newitt Rd	50 A2
Pankhurst Rd	50 A1
Pottery Rd	50 A1
Ratcliffe Highway	50 A1
Robson Dri	50 A1
Rochester Cres	50 A1
Ropers La	50 B1
St Johns Rd	50 A1
St Werburgh Cres	50 A2
Stoke Rd	50 B2
Sturdee Cotts	50 A2
The Bungalows	50 A2
Trubridge Rd	50 A2
Vicarage La	50 B2

Vidgeon Av	50 A1
Wall Clo	50 A1
Walters Rd	50 A1
Webb Clo	50 A1
Whitehouse Clo	50 B2
Wylie Rd	50 A2

HORSMONDEN

Angley Ct	50 B4
Back La	50 B4
Brenchley Rd	50 A4
Fromandez Dri	50 A4
Furnace La	50 A3
Gibbet La	50 A3
Goudhurst Rd	50 B4
Gunback La	50 A4
Gunlands	50 B3
Hoath Meadow	50 B4
Lamberhurst Rd	50 A4
Maidstone Rd	50 B4
Oast Vw	50 B4
Orchard Clo	50 A4
Orchard Cres	50 B3
Orchard Way	50 B4
The Green	50 B4

HYTHE

Albert La	51 A3
Albert Rd	51 A3
Albion Pl	51 B2
Arthur Rd	51 A3
Bank St	51 A2
Bartholomew St	51 A2
Basset Clo	51 B1
Basset Gdns	51 B1
Beaconsfield Ter	51 B3
Blackhouse Hill	51 B2
Blackhouse Rise	51 B1
Cannongate Av	51 B2
Cannongate Clo	51 B2
Cannongate Gdns	51 B2
Cannongate Rd	51 B2
Castle Av	51 A2
Castle Rd	51 A1
Chapel St	51 A2
Church Hill	51 A2
Church Rd	51 A2
Cliff Clo	51 B2
Cliff Rd	51 B1
Coastguard Cotts	51 A3
Cobay Clo	51 B2
Cobbs Pass	51 A2
Cobden Rd	51 A3
Deedes Clo	51 A1
Dental St	51 A2
Douglas Av	51 A2
Earlsfield Rd	51 B2
East St	51 A2
Elizabeth Gdns	51 A3
Elm Gdns	51 B2
Elm Pass	51 A2
Farmer Clo	51 B1
Fisher Clo	51 B3
Foys Pass	51 A2
Grange Rd	51 A1
Great Conduit St	51 A2
Hafod Pass	51 A2
High St	51 A2
Hillcrest Rd	51 A2
Hillside St	51 A2
Ladies Walk	51 A3
Lookers La	51 A1
Lower Blackhouse Hill	51 B2
Lucys Walk	51 A3
Lynton Rd	51 A3
Marine Par	51 A3
Marine Walk St	51 A2
Market Hill	51 A2
Mill La	51 B2
Mill Rd	51 A2
Mount St	51 A2
Moyle Tower Rd	51 A3
Napier Gdns	51 A3
New Rd	51 A3
Newington Meadow	51 B2
North Rd	51 A2
Oak Hall Pass	51 A2
Oak Walk	51 A2
Orchard Dri	51 A3
Ormonde Rd	51 A3
Park Rd	51 A3
Portland Rd	51 A2
Princes Par	51 B3
Prospect Rd	51 A2
Quarry Clo	51 A2
Quarry La	51 A2
Queens Ct	51 A3
St Hildas Rd	51 A3
St Leonards Rd	51 A3

School Rd	51 A1
Seabrook Rd	51 B2
Seaton Av	51 A2
Sene Park	51 B2
Sir William Pitt Clo	51 A2
Slade St	51 A3
South Rd	51 A3
Station Rd	51 A2
Sturdy Clo	51 B2
Sun La	51 A2
Tanners Hill	51 A1
Tanners Hill Gdns	51 A1
The Avenue	51 A2
The Close	51 A1
The Dene	51 A2
The Fairway	51 A3
Theatre St	51 A2
Theresa Rd	51 A3
Tower Gdns	51 A3
Twiss Av	51 A2
Twiss Gro	51 B2
Twiss Rd	51 B2
Upper Malthouse Hill	51 A2
Victoria Rd	51 A3
Wakefield Walk	51 A3
West Par	51 A3
Windmill St	51 A3
Wood Rd	51 A3

IGHTAM

Bates Hill	52 B2
Borough Green Rd	52 B2
Busty La	52 B2
Cobbs Clo	52 A2
Fen Pond Rd	52 A1
Ightham By-Pass	52 A2
Ightham Rd	52 A2
Jubilee Cres	52 A2
Mill La	52 B2
Oldbury Clo	52 A2
Oldbury Cotts	52 A2
Oldbury La	52 A2
Rectory La	52 A2
Spring La	52 A2
Stangate Quarry Rd	52 B1
The Close	52 B1
The Street	52 B2
Thong La	52 B2
Trycewell La	52 B2
Upper Spring La	52 A2

ISLE OF GRAIN

Chapel Rd	52 B4
Coronation Rd	52 B4
Doggetts Row	52 B4
Edinburgh Rd	52 B4
Fry Clo	52 B4
Grain Rd	52 A4
Grayne Av	52 B4
Green La	52 B4
High St	52 B4
Lapwing Rd	52 B4
Levet Clo	52 B3
Old Guard House	52 B4
Pannell Rd	52 B4
Pintail Clo	52 B4
Port Victoria Rd	52 B4
Puffin Rd	52 B4
St James Clo	52 B4
St James Rd	52 B4
Seaview	52 B4
Shelldrake Clo	52 B4
Smithfield Rd	52 B4
Teal Clo	52 B4
West La	52 A3

KEMSING

Barnfield Cres	53 A2
Beechy Lees Rd	53 A2
Birchin Cross Rd	53 A1
Boleyn Rd	53 A2
Brookfield	53 A2
Castle Dri	53 A2
Chart Vw	53 B2
Childsbridge La	53 A3
Church La	53 B2
Clarkes Green Rd	53 B1
Cleeves Rd	53 A2
Collet Rd	53 A2
Copperfield Walk	53 A2
Copperfields	53 A2
Copperfields Orchard	53 B2
Dippers Clo	53 B3
Dynes Rd	53 A2
Edgar Rd	53 A2
Gabriels Hill	53 A2
Greenlands Rd	53 B3

Greystones Clo	53 A2
Heaverham Rd	53 B2
High St	53 B2
Highfield Rd	53 A2
Hillside Rd	53 A2
Knave Wood Rd	53 A2
Mary Burrows Gdns	53 B2
Montfort Rd	53 A2
Nightingale Rd	53 A2
Noahs Ark	53 B3
Norman Clo	53 A2
Northdown Rd	53 A2
Orchard Way	53 B2
Oxenhill Rd	53 A2
Park Hill Rd	53 A2
Park La	53 B3
Pilgrims Way	53 A2
Rowdow La	53 A2
Rushymead	53 B3
St Ediths Rd	53 B2
St Michaels Dri	53 A2
Shorehill La	53 B1
Spring Head Rd	53 A2
The Chase	53 A2
The Dines	53 A2
The Landway	53 B2
The Parade	53 A2
West End	53 A2
Wulfred Way	53 B3

KINGSDOWN

Alexandra Rd	54 A2
Balmoral Rd	54 A2
Bayview Rd	54 A3
Boundary Rd	54 B2
Carlton Rd	54 A2
Cecil Rd	54 B1
Church Cliff	54 B2
Claremont Rd	54 A2
Cliff Rd	54 B2
Courtlands	54 B2
Edward Rd	54 A2
Glen Rd	54 A2
Granville Rd	54 B3
Hawksdown	54 A1
Hillcrest Rd	54 A3
Jarvist Pl	54 B2
Kings Clo	54 A2
Kingsdown Hill	54 A2
Kingsdown Rd	54 B1
North Rd	54 B2
Northcote Rd	54 B3
Oldstairs Rd	54 B3
Osborne Rd	54 A2
Queensdown Rd	54 A3
Ringwould Rd	54 A2
St James Rd	54 B2
St Monicas Rd	54 B2
Sea Rd	54 A2
South Rd	54 B2
The Avenue	54 A2
The Rise	54 A2
Underclifffe Rd	54 B2
Upper St	54 A2
Victoria Rd	54 A3
Wellington Par	54 B1

KINGSWOOD/
ULCOMBE

Ashford Dri	55 A1
Bell Way	55 A1
Broomfield Rd	55 A1
Bushy Gro	55 A1
Cayser Dri	55 A1
Chagworth Rd	55 B1
Charlesford Av	55 A1
Charlton La	55 A2
Chartway St	55 A1
Chestnut Clo	55 B3
Chestnut Dri	55 A1
Elder Clo	55 A1
Gravelly Bottom Rd	55 A1
Heather Wood Clo	55 A1
Holly Tree Clo	55 A1
Ivy Clo	55 A1
Laurel Clo	55 A1
Lenham Rd	55 B1
Lodge Gdns	55 B3
Morry La	55 A3
Streetfield	55 B3
Tall Tree Clo	55 A1
The Street	55 B3
The Wardens	55 A1
Tithe Walk	55 A1
Ulcombe Hill	55 B2
Water La	55 B1
Whitehall Dri	55 A1
Wildwood Clo	55 B1

Orchard Pl	59 A3	Allens	60 B2
Orchard St	59 A3	Ballards Clo	60 A1
Pads Hill	59 B2	Barrel Arch Clo	60 A1
Padsole La	59 B2	Bramley Ct	60 A1
Palace Av	59 A2	Chantry Pl	60 A2
Park Av	59 B1	Chantry Rd	60 A2
Peel St	59 A1	Church Grn	60 A1
Penenden St	59 A1	Copper La	60 B2
Perry St	59 A1	Cranham Sq	60 A2
Perryfield St	59 A1	Goudhurst Rd	60 A2
Pine Gro	59 B1	Haffenden Clo	60 A1
Postley Rd	59 A3	High St	60 A1
Princes St	59 B2	Howland Rd	60 B2
Priory Rd	59 A3	Jewell Gro	60 B2
Prospect Pl	59 A2	Lime Clo	60 A1
Pudding La	59 A2	Lucks Way	60 A1
Quarry Rd	59 A3	Maidstone Rd	60 B1
Quarry Sq	59 A1	Maplesden	60 B2
Queen Anne Rd	59 B2	Maynards	60 A2
Randall St	59 A1	Meades Clo	60 A1
Rawdon Rd	59 B3	Meadow Way	60 B1
Recreation Clo	59 B1	Oak Tree Clo	60 B2
Redcliffe La	59 B1	Pattenden La	60 A1
Reginald Rd	59 A3	Plain Rd	60 A2
Rock Rd	59 B1	South Rd	60 B2
Rocky Hill Ter	59 A2	Sovereigns Way	60 A1
Roman Hts	59 B1	Stanley Way	60 B2
Romney Pl	59 B2	Sutton Ct	60 A2
Rose Yd	59 A2	The Cockpit	60 A2
Royal Star Arcade	59 A2		

MARGATE

St Annes Ct	59 A2	Addington Rd	61 B1
St Lukes Av	59 B2	Addington Sq	61 B1
St Lukes Rd	59 B1	Addington St	61 B2
St Peter St	59 A2	Addiscombe Gdns	61 B2
St Phillips Av	59 B3	Addiscombe Rd	61 B2
St Stephens Sq	59 A3	Albert Rd	61 A2
Salem St	59 B3	Albert Ter	61 B1
Salisbury Rd	59 B1	Alexandra Homes	61 B2
Sandling Rd	59 A1	Alexandra Rd	61 B2
Saxons Dri	59 B1	Alexandra Ter	61 B2
Scott St	59 A1	Alkali Row	61 B1
Session House Sq	59 B2	All Saints Av	61 A2
Sheals Cres	59 A3	Alma Rd	61 B2
Shipley Ct	59 B2	Argyle Av	61 A2
Silchester Ct	59 B1	Argyle Gdns	61 A2
Sittingbourne Rd	59 B2	Arnold Rd	61 B2
Snowdon Av	59 B2	Athelstan Rd	61 B1
Snowdon Par	59 B2	Bath Pl	61 B1
Sportsfield	59 B1	Bath Rd	61 B1
Springfield Av	59 A1	Beatrice Rd	61 B2
Square Hill	59 B2	Belgrave Rd	61 B1
Square Hill Rd	59 B2	Bilton Sq	61 B1
Staceys St	59 A1	Broad St	61 B1
Stanhope Clo	59 A1	Brockly Rd	61 B1
Station Rd	59 A2	Buckingham Rd	61 B2
Stoneborough Centre		Buenos Ayres	61 A2
(Chequers)	59 A2	Burlington Gdns	61 A2
Stonebridge Centre	59 B2	Byron Rd	61 B2
Stuart Clo	59 B1	Canterbury Rd	61 A2
Sunningdale Ct	59 B2	Caroline Sq	61 B1
Terrace Rd	59 A2	Carroways Pl	61 B1
Terry Yard	59 B2	Caxton Rd	61 A2
The Mews	59 A2	Cecil Sq	61 B1
The Spinney	59 B3	Cecil St	61 B1
Thornhill Pl	59 A1	Chapel Hill Clo	61 B3
Tonbridge Rd	59 A3	Charlotte Sq	61 B1
Tovil Hill	59 A3	Church Rd	61 B2
Tovil Rd	59 A3	Church St	61 B2
Tudor Av	59 B1	Churchfields	61 B2
Tufton St	59 B2	Churchfields Pl	61 B1
Turkey St	59 B2	Cliff Ter	61 B1
Underwood Clo	59 A4	Clifton Gdns	61 B1
Union St	59 A2	Clifton Pl	61 B1
Upper Rd	59 B3	Clifton Rd	61 B1
Upper Stone St	59 B3	Clifton St	61 B1
Vinters Rd	59 B2	Cobbs Pl	61 B1
Water La	59 A2	College Rd	61 B1
Waterloo Rd	59 B1	Connaught Gdns	61 B2
Waterloo St	59 B3	Connaught Rd	61 B2
Waterside	59 A2	Cowper Rd	61 B2
Week St	59 A2	Craven Clo	61 A3
Well Rd	59 A1	Crescent Rd	61 A2
Wellington Pl	59 A1	Dane Hill	61 B1
West Park Rd	59 B3	Dane Hill Row	61 B1
Westree Rd	59 A3	Dane Mead Ter	61 B1
Whatman Clo	59 B1	Dane Park Rd	61 B1
Wheeler St	59 B2	Dane Rd	61 B1
Wheeler St Hedges	59 B1	Darlby Rd	61 B1
Whiterock Pl	59 A2	Darlby Sq	61 B1
Willow Way	59 B3	Drapers Av	61 B1
Winchester House	59 B1	Duke St	61 B1
Windsor Clo	59 B1	Durban Rd	61 B1
Woodville Rd	59 B3	Eaton Hill	61 B1
Woollett St	59 A2	Eaton Rd	61 B1
Wyatt St	59 B2	Elmley Way	61 B3
Wyke Manor Rd	59 A2	Empire Ter	61 B3
York Rd	59 B3	Ethelbert Gdns	61 B1
		Ethelbert Rd	61 B1
		Ethelbert Ter	61 B1

MARDEN

Albion Rd	60 B2

Farley Rd	61 B3	Westbrook Gdns	61 A2
Firbank Gdns	61 A3	Westbrook Prom	61 A2
Fort Cres	61 B1	Westbrook Rd	61 A2
Fort Hill	61 B1	Westcliff Rd	61 A2
Fort Prom	61 B1	Westfield Rd	61 A2
Fort Rd	61 B1	Yoakley Sq	61 B2
Fulham Av	61 A3	Zion Pl	61 B1
Garfield Rd	61 A2		

MEOPHAM

George V Av	61 A2	Arnold Rd	62 B3
Giles Gdns	61 B2	Birtrick Dri	62 A1
Gladstone Rd	61 B2	Blenheim Clo	62 B3
Grosvenor Gdns	61 B2	Camer Park Rd	62 B2
Grosvenor Hill	61 B2	Camer Rd	62 B2
Grosvenor Pl	61 B1	Camer St	62 B2
Grotto Gdns	61 B1	Cedar Clo	62 B2
Grotto Hill	61 B1	Chestnut Clo	62 A2
Grotto Rd	61 B1	Cheyne Walk	62 B3
Grove Gdns	61 A2	Copt Hall Rd	62 B1
Halfmile Ride	61 B3	Denesway	62 B2
Hartsdown Rd	61 A2	Dormers Dri	62 B2
Hatfield Rd	61 A2	Ediva Rd	62 B1
Hawley Sq	61 B1	Edmund Clo	62 B1
Hawley St	61 B1	Evenden Rd	62 A2
Helens Av	61 B2	Fairview Gdns	62 B1
High St	61 B1	Foxendown Rd	62 B3
Highfield Gdns	61 A2	Green La	62 B2
Jacob Clo	61 B3	Grenville Clo	62 B3
King St	61 B1	Hadley Clo	62 B3
Kingston Av	61 A3	Huntingfield Rd	62 B2
Knold Park	61 A3	Johns Rd	62 A1
Lausanne Rd	61 B1	Lances Clo	62 B2
Lausanne Ter	61 B2	Lilac Pl	62 B2
Lister Rd	61 B2	Longfield Rd	62 A2
Lombard St	61 B1	Meadfield Rd	62 B3
Love La	61 B1	Melliker La	62 A2
Manston Rd	61 A3	Mulberry Clo	62 B2
Marine Dri	61 B1	New Rd	62 A1
Marine Gdns	61 B1	Norwood La	62 B1
Marine Ter	61 A1	Nursery Rd	62 B1
Market St	61 B1	Oakmead	62 B3
Marlborough Rd	61 B2	Orchard Way	62 A1
Mill La	61 B2	Park Hill	62 A1
Milton Av	61 B2	Pine Rise	62 B2
Nash Court Gdns	61 B2	Poplar Way	62 B2
Nash Court Rd	61 B3	Rowan Clo	62 B2
Nash La	61 B3	School Clo	62 B2
Nash Rd	61 B3	Shipley Hills Rd	62 A3
New Cross St	61 B1	Station Rd	62 A1
New St	61 B1	Strand Clo	62 B2
Northdown Rd	61 B1	The Medlars	62 A2
Oast Ct	61 B2	The Pippins	62 A2
Osborne Ter	61 B2	The Russetts	62 A2
Oxford St	61 B2	The Street	62 B3
Park Cres Rd	61 B2	Tradescant Way	62 B2
Park La	61 B2	Walnut Tree Way	62 B2
Park Pl	61 B2	Warwick Gdns	62 B3
Park Rd	61 B2	White Post La	62 B1
Payton Clo	61 B3	Whitehill Rd	62 B3
Perkins Av	61 B2	Wrotham Rd	62 A1
Poets Corner	61 B2		

MINSTER-IN-SHEPPEY

Princess Cres	61 B2	Abbey Clo	63 B2
Princess St	61 B1	Abbeyview Dri	63 A2
Queens Av	61 B2	Augustine Rd	63 A1
Railway Ter	61 A2	Back La	63 B2
Ramsgate Rd	61 B2	Baldwin Rd	63 B2
Rancorn Rd	61 A2	Barton Hill Dri	63 A3
Royal Cres	61 A2	Bellvue Rd	63 A2
Royal Esp	61 A2	Blatcher Clo	63 A2
St Andrews Clo	61 B3	Boundary Clo	63 B2
St Annes Gdns	61 B3	Bramston Rd	63 A2
St Augustines Av	61 B2	Brecon Chase	63 A2
St Johns Rd	61 B2	Broadway	63 A1
St Johns St	61 B2	Chapel St	63 B2
St Peters Footpath	61 B2	Chiddingford Clo	63 B2
St Peters Rd	61 B2	Church Rd	63 B2
Salmestone Rd	61 B2	Cliff Gdns	63 B2
Sea View Ter	61 A2	Clovelly Dri	63 A1
Setterfield Rd	61 B2	Copland Av	63 A2
Shakespeare Rd	61 A2	Darlington Dri	63 A2
Shottendane Rd	61 A3	Drake Av	63 B2
Station Rd	61 A2	Dreadnought Av	63 A2
Sussex Av	61 B2	Echo Way	63 B2
Thanet Rd	61 B1	Edwina Av	63 A2
The Parade	61 B1	Elm La	63 B3
The Rendezvous	61 B1	Emley Rd	63 B3
Tivoli Park Av	61 A2	Falcon Gdns	63 B2
Tivoli Rd	61 A2	Fleetwood Clo	63 A2
Trinity Pl	61 B1	Forty Acres Hill	63 A3
Trinity Sq	61 B1	Glendale Rd	63 A1
Ulster Rd	61 B2	Glenwood Dri	63 A2
Union Cres	61 B1	Greyhurst Rd	63 B3
Union Row	61 B1	Harps Av	63 A2
Upper Gro	61 B1	Harps Walk	63 A2
Vicarage Cres	61 B2	High St	63 B2
Vicarage Pl	61 B2	Highview Rd	63 B2
Victoria Rd	61 B2		
Walpole Rd	61 B1		
Waverley Rd	61 A2		
Wellis Gdns	61 A2		
Westbrook Av	61 A2		
Westbrook Cotts	61 B3		

Hillside Rd	63 A2	Ayelands Av	65 A2
Howard Clo	63 A1	Bazes Shaw	65 B2
Imperial Av	63 B2	Billet Hill	65 A3
Johnson Way	63 A2	Bowes Wood	65 B2
Kent Av	63 A2	Butchers La	65 A2
Kings Rd	63 B2	Caling Croft	65 B1
Love La	63 B2	Capelands	65 B2
Lower Rd	63 A3	Centre Rd	65 A2
Lynmouth Dri	63 B2	Chapel Wood	65 A2
Marina Dri	63 A2	Chapel Wood Rd	65 A2
Mill Hill	63 B2	Church Rd	65 A2
Minster Dri	63 A1	Church Rd, Ash	65 A3
Minster Rd	63 A2	Colt Stead	65 A2
Nautilus Clo	63 A2	Farm Holt	65 B1
Nautilus Dri	63 A2	Foxbury	65 A2
Nelson Av	63 B2	Grange La	65 B1
New Rd	63 A2	Hartley Bottom Rd	65 B3
Noreen Av	63 A2	Knights Croft	65 A3
Norwood Rise	63 A2	Knightscroft Rd	65 B3
Orchard Gro	63 A2	Lambardes	65 B2
Parish Rd	63 A3	Manor Forstal	65 B2
Petfield Clo	63 B2	Millfield La	65 A2
Porter Clo	63 A2	Millfield	65 A2
Prince Charles Av	63 B2	North Ash Rd	65 A2
Princess Av	63 B2	Olivers Mill	65 A2
Queenborough Dri	63 A2	Over Minnis	65 B2
Queens Rd	63 B2	Pease Hill	65 A3
Saxon Av	63 A2	Penenden	65 A2
Scarborough Dri	63 A1	Punch Croft	65 A2
Scrapsgate Rd	63 A2	Redhill Rd	65 A3
Sea Side Av	63 A1	Redhill Wood	65 B2
Seathorpe	63 B2	Seven Acres	65 A2
Sexburga Dri	63 A1	South Ash Rd	65 A3
Shurland Av	63 A2	Spring Cross	65 B2
Southsea Av	63 A1	The Street	65 A3
Stanley Av	63 B2	Upper North St	65 A2
Sunnyside Av	63 A2	West Yoke Rd	65 A2
Tams Gdns	63 B2	Westfield	65 B3
The Broadway	63 A1		
The Glen	63 A1		
The Leas	63 A1		

NEWINGTON

The Maples	63 A2	Allsworth Clo	67 A2
The Rowans	63 A2	Boxted La	67 A1
Union Rd	63 B2	Boyces Hill	67 B2
Vicarage Rd	63 B2	Bramley Clo	67 A2
Wards Hill Rd	63 A1	Brooks Pl	67 A2
Waterloo Hill	63 B2	Bull La	67 A2
Waverley Av	63 A2	Callaways La	67 A2
Westcliff Dri	63 B1	Church La	67 A2
Whitethorne Gdns	63 B2	Denham Rd	67 A1
Whybornes Chase	63 A2	Hasted Rd	67 A1
Windmill Rise	63 B2	High Oak Hill	67 B1
Woodland Dri	63 A1	High St	67 A2
Worcester Clo	63 A2	London Rd	67 A1
		Orchard Dri	67 A2
		Pear Tree Walk	67 A2
		Playstool Clo	67 A2
		Playstool Rd	67 A2
		St Marks Clo	67 B1

MINSTER-IN-THANET

Abbey Gro	64 B3	St Martins Clo	67 B1
Augustine Rd	64 A2	St Marys Vw	67 B1
Bedlam Court La	64 B3	St Mathews Clo	67 B1
Brockmans Clo	64 A2	St Stephens Clo	67 B1
Canterbury Rd West	64 A1	School La	67 A1
Cheesmans Clo	64 B3	Station Rd	67 A2
Church St	64 B3	The Tracies	67 A2
Conyngham Rd	64 B3	The Willows	67 A2
Domneva Rd	64 A2	Vicarage Ct	67 A1
Durlock	64 B3	Westwood Clo	67 A1
Edgar Rd	64 A2	Wickham Clo	67 A2

NEW ROMNEY

Egbert Rd	64 A2	Ashdown Clo	66 A2
Fairfield Rd	64 B2	Ashford Rd	66 A1
Foxborough La	64 B2	Blenheim Rd	66 C2
Freemans Rd	64 A2	Brissenden Clo	66 B1
Greenhill Clo	64 A2	Broadlands Av	66 B1
Greenhill Gdns	64 A2	Broadlands Cres	66 B1
High St	64 B2	Cannon St	66 B1
Kenton Gdns	64 A2	Cherry Gdns	66 C2
Laundry Rd	64 B2	Church App	66 A2
Marsh Farm Rd	64 B3	Church La	66 A2
Molineux Rd	64 A2	Church Rd	66 A2
Monkton Rd	64 A2	Clarendon Mews	66 B1
Norton Dri	64 B2	Coast Rd	66 C2
Orchard Clo	64 B2	Cockreed La	66 A1
Petts Cres	64 A2	Craythorne Clo	66 B1
Prospect Gdns	64 A2	Craythorne La	66 B1
Prospect Rd	64 A2	Darcy Sq	66 C2
Rose Gdns	64 A2	Ellesmere Mews	66 B1
St Marys Rd	64 B2	Fairfield Clo	66 A2
St Mildreds Rd	64 B3	Fairfield Rd	66 A2
Station App	64 B3	George La	66 B2
Station Rd	64 B3	Grand Par	66 C2
Taylor Rd	64 A2	Hardwick Dri	66 B1
Thorne Rd	64 A2	Haywards Clo	66 B2
Tothill St	64 B2	High St	66 A2
Watchester La	64 B3	Langport Rd	66 B2
		Links Way	66 C2

NEW ASH GREEN

Ash Rd	65 A1	Littlestone Rd	66 B2
Ayelands	65 A2		

Lydd Rd	66 A2	Aynscombe Angle	68 B2
Lyons Rd	66 A2	Badgers Copse	68 A3
Mabledon Clo	66 A2	Bancroft Gdns	68 A2
Madeira Rd	66 C2	Bank Hart Rd	68 B2
Marine Par	66 C2	Barnsdale Cres	68 B1
Marlborough Clo	66 C2	Barry Clo	68 A3
Marsh Cres	66 B2	Becketts Clo	68 A3
Mountfield Rd	66 B2	Bedford Rd	68 B2
North St	66 A2	Bellefield Rd	68 B1
Oak Lodge Rd	66 B1	Benedict Clo	68 A3
Park Rd	66 C2	Berkeley Clo	68 A2
Pembroke Mews	66 B1	Berrylands	68 B3
Priory Clo	66 A2	Berwick Way	68 A2
Queens Rd	66 C2	Bicknor Rd	68 A2
Richmond Dri	66 B1	Bishop Butt Clo	68 A3
Rolfe La	66 A2	Blenheim Rd	68 B3
Ryswick Mews	66 B1	Bosco Clo	68 A3
St Andrews Rd	66 C2	Bournewood Rd	68 B1
St Johns Rd	66 B2	Brasted Clo	68 A3
St Martins Rd	66 B2	Bridge Rd	68 B1
St Marys Rd	66 B2	Brookmead Clo	68 B1
St Nicholas Rd	66 B2	Brookmead Way	68 B1
Spitalfield La	66 A2	Brookside	68 A2
Station App	66 B2	Broomhill Rd	68 A2
Station Rd	66 A2	Broxbourne Rd	68 A2
Sussex Rd	66 A2	Bruce Gro	68 B2
The Churchlands	66 B2	Buckingham Clo	68 A2
The Fairway	66 C2	Buckland Rd	68 A2
Tookey Rd	66 A3	Cambray Rd	68 A2
Tritton La	66 A2	Carlton Par	68 B2
Victoria Rd	66 C2	Carolyn Dri	68 B3
Victoria Rd West	66 C2	Cathcart Dri	68 A2
Victoria St	66 A2	Cedars Rd	68 A2
Walner Gdns	66 B1	Charterhouse Rd	68 A3
Walner La	66 B1	Chelsfield Rd	68 B1
Warren Rd	66 B2	Cheltenham Rd	68 A3
West St	66 A2	Chislehurst Rd	68 A1
Wiles Av	66 B2	Church Hill	68 A1
Windsor Mews	66 B1	Church Hill Wood	68 A1
		Clovelly Way	68 A1
		Coleridge Way	68 B1

NORTHFLEET

Bankside	67 A4	Cornmill Dri	68 A2
Burnaby Rd	67 B4	Cotswold Rise	68 A1
College Rd	67 A3	Court Rd	68 B2
Council Av	67 A4	Cowden Rd	68 A2
Crete Hall Rd	67 B3	Cranleigh Clo	68 A3
Dover Rd	67 B4	Cray Av	68 B1
East Kent Av	67 A4	Crofton La	68 A2
Ebbsfleet Walk	67 A3	Crofton Rd	68 A2
Factory Rd	67 A3	Cromford Rd	68 A3
Fishermans Hill	67 A3	Cross Rd	68 B1
Ford Rd	67 A3	Cyril Rd	68 A2
Gordon Rd	67 B4	Dale Wood Rd	68 A2
Granby Rd	67 B4	Dalton Clo	68 A3
Grove Rd	67 A3	Dene Dri	68 B3
Hamerton Rd	67 A3	Denver Clo	68 A1
Hartfield Pl	67 B4	Devonshire Rd	68 A2
High St	67 A3	Dorley Gdns	68 B3
Hive La	67 A3	Dryden Way	68 B2
Huntley Av	67 A4	Dyke Dri	68 B2
Kent Av	67 A4	East Dri	68 B1
Kingston Ct	67 A3	Eastcote	68 A2
Laburnum Gro	67 B4	Edmund Rd	68 B1
Lawn Rd	67 A3	Elm Gro	68 A2
Lime Av	67 B4	Elmcroft Rd	68 B2
London Rd	67 B4	Elysian Rd	68 A1
Masefield Rd	67 B3	Felstead Rd	68 B3
Mill Rd	67 B4	Footbury Hill Rd	68 B2
Plane Av	67 B4	Fordcroft Rd	68 B1
Portland Rd	67 B3	Fordwich Clo	68 A2
Railway St	67 A3	Forest Way	68 A1
Robinia Av	67 B4	Friar Rd	68 A1
Rose St	67 A3	Gillmans Rd	68 B2
Rural Vale	67 B4	Gilroy Way	68 B1
Shepherd St	67 B4	Glendower Cres	68 B1
Springhead Rd	67 B4	Goddington La	68 A3
Station Rd	67 A3	Godington Chase	68 B3
Stonebridge Rd	67 A3	Goodmead Rd	68 B1
The Creek	67 A3	Grampian Clo	68 A1
The Hill	67 B4	Gravel Pit Way	68 A3
The Shore	67 B3	Greencourt Rd	68 A1
Tooley St	67 B4	Greenwood Clo	68 A1
Vicarage Dri	67 A3	Grosvenor Rd	68 A1
Wallis Pk	67 A3	Haydens Clo	68 B1
Warwick Pl	67 A3	Hayfield Rd	68 B1
West Kent Av	67 A4	Healy Dri	68 A3
York Rd	67 B4	Heath Clo	68 B1
		High St	68 A2

ORPINGTON

Albert Rd	68 B1	High St, St Mary Cray	68 B1
Ambrose Clo	68 A3	Highlands Rd	68 B2
Anglesea Rd	68 B1	Hill End	68 A3
Archer Rd	68 A1	Hill View Cres	68 A2
Arne Gro	68 A3	Hill View Rd	68 A3
Audley Walk	68 B1	Hillcrest Rd	68 A3
Augustus La	68 A2	Homefield Rise	68 B2
Austin Rd	68 B1	Hood Av	68 B1
Avalon Rd	68 B3	Horsecroft Clo	68 B2
Aylesham Rd	68 A1	Horsmonden Clo	68 A2
		Irene Rd	68 A2
		Irvine Way	68 A2

Juglands Rd 68 B2
Kedleston Dri 68 A1
Kent Rd 68 B1
Keswick Rd 68 A2
Knoll Rise 68 A2
Kynaston Rd 68 B2
Ladywood Rd 68 A1
Lagoon Rd 68 B1
Lamorna Clo 68 A2
Lancing Rd 68 B2
Lapworth Clo 68 B3
Larkspur Clo 68 B3
Leamington Av 68 A3
Lincoln Green Rd 68 A1
Littlejohn Rd 68 B1
Lockesley Dri 68 A1
Lodge Clo 68 B2
Lodge Cres 68 B2
Lower Rd 68 B1
Lucerne Rd 68 A2
Lychgate Rd 68 B2
Lynwood Gro 68 A2
Maltby Clo 68 A2
Marcellina Way 68 A3
Marion Cres 68 A1
Marlborough Clo 68 A1
Marsden Way 68 A3
Maxwell Gdns 68 A3
May Av 68 B1
Mayfield Av 68 A2
Meath Clo 68 B1
Melbourne Clo 68 A2
Mimosa Clo 68 B3
Mitchell Rd 68 A3
Moorfield Rd 68 B2
Mortimer Rd 68 B2
Mountview 68 A2
New Rd 68 B2
Newstead Av 68 A3
Northfield Rd 68 B1
Notholme Rise 68 A3
Novar Clo 68 A2
Nursery Clo 68 A2
Oakdene Rd 68 A1
Oakfield Rd 68 B2
Oakhill Rd 68 A2
Oatfield Rd 68 A2
Orchard Grn 68 A3
Orchard Gro 68 A2
Orchard Rd 68 B2
Ottenden Clo 68 A3
Packham Clo 68 B3
Park Av 68 A3
Park Rd 68 B1
Pendennis Rd 68 B2
Pepys Rise 68 A2
Perry Hall Clo 68 B1
Perry Hall Rd 68 A1
Petts Wood Rd 68 A1
Polperro Clo 68 A1
Pondwood Rise 68 A2
Poverest Rd 68 A1
Poynings Clo 68 B3
Princes Av 68 A1
Ramsden Rd 68 B2
Red Lion Clo 68 B1
Repton Rd 68 A3
Ridgeway Cres 68 A3
Ridgeway Cres Gdns 68 A3
Robin Hood Grn 68 A1
Roseberry Gdns 68 A3
Rosecott Clo 68 B1
St Andrews Dri 68 B1
St Aubyns Gdns 68 A3
St Francis Clo 68 A1
St Johns Rd 68 A1
St Josephs Clo 68 A3
St Kilda Rd 68 A2
St Margarets Clo 68 B3
Sanderstead Rd 68 B1
Sandhurst Rd 68 A3
Sandy La 68 A2
Scadshill Clo 68 A1
Sequdia Gdns 68 A2
Sevenoaks Rd 68 A3
Shelley Clo 68 A3
Shepherds Clo 68 A3
Sherlies Rd 68 A3
Somerset Rd 68 B2
Southcroft Rd 68 A3
Southfleet Rd 68 A3
Spencer Clo 68 A3
Spur Rd 68 A3
Stanley Rd 68 A2
Stanley Way 68 B1
Stapleton Rd 68 A3
Station App 68 A2
Station Rd 68 A3
Stephen Clo 68 A3
Strickland Way 68 A3
Sussex Rd 68 B1

Sutherland Av 68 A1
Swan Rd 68 A1
Taylor Clo 68 A3
Telscombe Clo 68 A3
The Approach 68 A2
The Avenue 68 A3
The Chenies 68 A1
The Covert 68 A1
The Drive 68 A3
The Greenways 68 B1
The Maltings 68 A2
The Walnuts 68 B2
Thorpe Clo 68 A3
Tintagel Rd 68 B3
Tower Clo 68 A3
Tower Rd 68 A3
Tubbenden Clo 68 A3
Tubbenden La 68 A3
Uplands Rd 68 B2
Vinson Clo 68 B2
Walnuts Rd 68 B2
Warwick Clo 68 B3
Wayne Clo 68 A3
Wellington Rd 68 B1
Wendover Way 68 B1
Westholme 68 A2
White Hart Rd 68 B2
Willett Way 68 A1
Willow Clo 68 B1
Wiltshire Rd 68 A2
Woodhead Dri 68 A2
Woodley Rd 68 B3
Worth Clo 68 A3
Wye Clo 68 A2
Wyvern Clo 68 B3
Yeovil Clo 68 A3
Zelah Rd 68 B2

OTFORD

Broughton Rd 69 A2
Bubblestone 69 B2
Colets Orchard 69 B2
Coombe Rd 69 B1
Darnetts Field 69 A2
Evelyn Rd 69 B2
Flower Field 69 A2
Greenhill Rd 69 B1
High St 69 A2
Hillydeal Rd 69 B1
Knighton Rd 69 A2
Leonard Av 69 B2
Old Otford Rd 69 B3
Orchard Rd 69 A2
Otford Rd 69 B3
Pickmoss La 69 A2
Pilgrims Way East 69 B1
Pilgrims Way West 69 A2
Rye Croft 69 A2
Rye La 69 A2
Sevenoaks Rd 69 B2
Shoreham Rd 69 B1
Sidney Gdns 69 B2
Station Rd 69 B2
Telston La 69 A2
The Butts 69 B2
The Charne 69 A2
The Old Walk 69 B2
Tudor Cres 69 B2
Tudor Dri 69 B2
Vestry Rd 69 B3
Warham Rd 69 A2
Well Rd 69 B2
Willow Park 69 A2

PADDOCK WOOD

Alliance Way 70 A2
Allington Rd 70 A2
Ashcroft Rd 70 A3
Badsell Rd 70 A3
Ballard Way 70 B2
Birch Rd 70 A2
Bowls Pl 70 A2
Bramley Gdns 70 A2
Bullion Clo 70 A2
Catts Pl 70 B3
Challenger Clo 70 A2
Chantlers Hill 70 A3
Church Rd 70 A2
Claver Dell Rd 70 A2
Cobbs Clo 70 A2
Cogate Rd 70 A2
Commercial Rd 70 A2
Concord Clo 70 A2
Dimmock Clo 70 B2
Eastwell Clo 70 A2
Eldon Way 70 A2
Forest Rd 70 A2
Fuggles Clo 70 A3
Goldings 70 A3
Granary 70 B2

Haywain Clo 70 A3
Hop Pocket La 70 A2
Hornbeam Clo 70 A3
Kent Clo 70 A2
Keyworth Clo 70 A2
Kiln Way 70 A3
Larch Gro 70 A2
Laxton Gdns 70 A2
Le Temple Rd 70 B2
Linden Clo 70 A3
Lucknow Rd 70 A2
Lucks La 70 B1
MacDonald Ct 70 A2
Maidstone Rd 70 A1
Mascalls Court Rd 70 A3
Mascalls Park 70 A3
Mercers Clo 70 A2
Mile Oak Rd 70 B3
Mount Pleasant 70 A2
New Rd 70 A2
Newton Gdns 70 A2
North Down Clo 70 A3
Nursery Rd 70 A2
Oaklea Rd 70 A2
Old Kent Rd 70 A2
Pearsons Green Rd 70 B3
Pinewood Clo 70 B3
Queen St 70 B3
Ribston Gdns 70 A2
Ringden Av 70 A3
St Andrews Clo 70 A2
St Andrews Rd 70 A2
Station App 70 A2
Station Rd 70 A2
Sycamore Gdns 70 A3
The Bines 70 A3
The Greenways 70 A2
Tutsham Way 70 A2
Wagon La 70 B1
Walnut Clo 70 A2
Warrington Rd 70 A2
Woodlands 70 A2
Yeoman Gdns 70 A2

PEMBURY

Amberleaze Dri 71 B3
Batchelors 71 B2
Beagles Wood Rd 71 B2
Belfield Rd 71 B3
Bulls Pl 71 B3
Camden Av 71 A3
Camden Ct 71 B3
Canterbury Rd 71 B3
Chalket La 71 A3
Cornford Clo 71 A3
Cornford La 71 A3
Elmhurst Av 71 B2
Forest Vw 71 B2
Forest Way 71 B2
Gimble Way 71 B2
Greenleas 71 A3
Hastings Rd 71 B3
Henwood Green Rd 71 B2
Henwoods Cres 71 B3
Henwoods Mount 71 B3
Herons Way 71 B2
Heskett Pk 71 B3
High St 71 A3
Highfield Clo 71 B3
Knights Clo 71 B2
Knights Ridge 71 B2
Lower Green Rd 71 B2
Maidstone Rd 71 B2
Middle Field 71 B2
Old Church Rd 71 B1
Paddock Clo 71 A3
Pembury Clo 71 B2
Pembury Rd 71 A3
Pembury Walks 71 A1
Petersfield 71 B2
Redwings La 71 B1
Ridgeway 71 B2
Romford Rd 71 B2
Rowans 71 B3
Rowley Hill 71 B1
Sandhurst Av 71 B3
Snipe Clo 71 B2
Stanam Rd 71 B3
Stonecourt La 71 B2
Sunhill Clo 71 A3
Sweepshill Clo 71 B2
Sycamore Cotts 71 B2
The Coppice 71 B2
The Forstal 71 B2
The Gill 71 B2
The Glebe 71 B2
The Grove 71 B2
The Meadow 71 B2
The Paddock 71 A3
Tonbridge Rd 71 A2

Westway 71 B2
Woodgate Way 71 A3
Woodhill Pk 71 A3
Woodside Clo 71 B3
Woodside Rd 71 B3

QUEENBOROUGH

Alsager Rd 58 A4
Argent Rd 58 A4
Barler Rd 58 B3
Bartletts Clo 58 B3
Borough Rd 58 B3
Brielle Way 58 B3
Castle St 58 A3
Castlemere Av 58 B3
Chalk Rd 58 B3
Coronation Cres 58 A3
Court Hall 58 A3
Cullet Dri 58 A4
Dumergue Av 58 B3
Eastern Av 58 B3
Edward Rd 58 B3
Ferry Way 58 A4
First Av 58 A4
Foxley Rd 58 A3
Georgian Clo 58 B4
Gordon Av 58 B3
Harold St 58 B3
High St 58 A3
Hillside Av 58 A4
Jubilee Cres 58 A3
Lower Rd 58 B3
Main Rd 58 B3
Manor Rd 58 A4
Marshall Cres 58 A4
Moat Way 58 B3
North Rd 58 B3
Park Av 58 B3
Park Rd 58 A3
Queenborough Rd 58 B3
Railway Ter 58 A3
River Vw 58 A4
Rushenden Rd 58 A4
Second Av 58 A4
Sheet Glass Rd 58 A4
South St 58 A3
Stanley Av 58 B3
Sterling Rd 58 B3
Swale Av 58 A4
The Rise 58 B3
Uplands Way 58 B3
Well Rd 58 A3
Well Rd 58 A4
West St 58 A3
Wykeham Clo 58 A4
Yevele Clo 58 B3

RAINHAM

Arthur Rd 72 A2
Ashley Rd 72 A1
Asquith Rd 72 A3
Balmer Clo 72 A2
Barleycorn Dri 72 B3
Beacon Clo 72 A3
Bedford Av 72 A3
Beechings Grn 72 A1
Beechings Way 72 A1
Begonia Av 72 A1
Bendon Way 72 A2
Berengrave La 72 B2
Bettescombe Rd 72 A3
Beverley Clo 72 B2
Birling Av 72 A2
Blean Rd 72 A1
Bloors La 72 A2
Bodian Clo 72 A1
Bonnington Grn 72 A1
Boughton Clo 72 A1
Brabourne Av 72 A1
Bransgore Clo 72 A3
Broadview Av 72 B2
Brockenhurst Clo 72 A3
Broomcroft Rd 72 B1
Brown Rd 72 B2
Bushmeadow Rd 72 B1
Caldew Av 72 A2
Caledonian Ct 72 B2
Callans Clo 72 A3
Cambridge Rd 72 A3
Camellia Clo 72 A3
Century Rd 72 A2
Chalfont Dri 72 A3
Chalky Bank Rd 72 B1
Cheriton Rd 72 A2
Cherry Amber Clo 72 B2
Cherry Tree Rd 72 B2
Chesham Dri 72 A3
Chesterfield Clo 72 B1
Childscroft Rd 72 B1
Chilton Ct 72 B2

Church Mews	72 B2	Queendown Av	72 A3
Cowbeck Clo	72 A3	Quinnell St	72 B2
Cozenton Ct	72 B2	Ringwood Clo	72 A2
Crabtree Rd	72 A2	Ripon Clo	72 A1
Cranbrook Clo	72 A1	Roberts Rd	72 B2
Cranford Clo	72 A2	Rolvenden Av	72 A1
Crundale Rd	72 A1	Romany Rd	72 A1
Danson Way	72 A2	Roystons Clo	72 B1
De Mere Clo	72 A3	Ruckinge Way	72 A1
Deanwood Clo	72 A3	Salisbury Av	72 A2
Deanwood Dri	72 A3	Sandhurst Clo	72 A1
Denbigh Dri	72 A2	Sandown Dri	72 A3
Derwent Way	72 A2	Selinge Grn	72 A1
Detling Clo	72 A1	Selsted Clo	72 A2
Devon Clo	72 B2	Sheldon Dri	72 B2
Dignals Clo	72 B1	Signal Ct	72 B2
Doddington Rd	72 A1	Silverdale Rd	72 B3
Dorset Sq	72 A2	Silverspot Clo	72 B3
Durham Rd	72 A3	Soloman Rd	72 B3
Eastling Clo	72 A1	Springvale	72 A3
Edwin Rd	72 A2	Station Rd	72 B2
Elizabeth Ct	72 A1	Stratford Av	72 A2
Elmstone Rd	72 A3	Streetfield Rd	72 B2
Ely Clo	72 B1	Sturry Way	72 A1
Fordwich Grn	72 A1	Suffolk Av	72 B2
Frinstead Clo	72 A1	Suffolk Ct	72 B2
Gainsborough Clo	72 A3	Suffolk Dri	72 A2
Gayhurst Clo	72 A3	Sunningdale Clo	72 A3
Glistening Glade	72 B3	Sunningdale Dri	72 A3
Granary Clo	72 B2	Sunnyfield Clo	72 B2
Guardian Ct	72 A2	Sutherland Gdns	72 B3
Hartpiece Clo	72 B1	Sylvan Rd	72 A2
Harvesters Clo	72 B3	Tanker Hill	72 A3
Harvey Rd	72 A2	Taverners Rd	72 A3
Hawthorne Av	72 A1	Tavistock Clo	72 B3
Henley Clo	72 A1	Thames Av	72 B2
Herbert Rd	72 A2	The Goldings	72 A2
Hereford Clo	72 A1	The Mailyns	72 A3
Herne Rd	72 A1	The Platters	72 A3
High Elms	72 B1	The Willows	72 B1
High St	72 B2	Thornton Rd	72 A1
Highfield Clo	72 A3	Truro Rd	72 A1
Highfield Rd	72 A2	Tudor Gro	72 B2
Holding St	72 B2	Tufton Rd	72 B2
Hollingsbourne Rd	72 A1	Waltham Rd	72 A1
Holmoaks	72 B1	Waterworks La	72 A2
Holtwood Clo	72 A3	Webster Rd	72 B2
Honeybee Glade	72 B3	Wentworth Dri	72 A3
Hothfield Rd	72 B2	Wheatcroft Gro	72 B2
Hurst Pl	72 B2	Whitegate Ct	72 A3
Idenwood Clo	72 A3	Wimbourne Dri	72 A3
Inversgate Clo	72 B1	Windermere Dri	72 A3
Ivy St	72 B2	Wingham Clo	72 A1
Kendal Way	72 A2	Woodchurch Cres	72 A1
Kenilworth Dri	72 A3	Woodpecker Glade	72 A3
Kenilworth Gdns	72 A3	Woodside	72 A3
Lakewood Dri	72 A8	Wooldeys Rd	72 B1
Lancaster Ct	72 A3	Wooton Grn	72 A1
Langdale Clo	72 A2		
Laurel Walk	72 B3		

RAMSGATE

Lichfield Clo	72 A1	Abbotts Hill	73 A3
Lineacre Clo	72 A3	Addington Pl	73 A3
London Rd	72 A2	Addington St	73 A3
Longford Clo	72 B2	Adelaide Gdns	73 A3
Longley Rd	72 B2	Albert Rd	73 B2
Lonsdale Dri	72 A3	Albert St	73 A3
Lower Rainham Rd	72 B1	Albion Hill	73 A3
Lyminge Clo	72 A1	Albion Pl	73 A3
Lyndhurst Av	72 A3	Albion Rd	73 B2
Maidstone Rd	72 A3	Alexandra Rd	73 A2
Maplins Clo	72 B2	Alma Pl	73 A2
Marshall Rd	72 A2	Alma Rd	73 A2
Mayfield Clo	72 B1	Alpha Rd	73 A2
Medway Rd	72 B2	Anns Rd	73 A2
Megby Clo	72 A3	Archway Rd	73 A3
Meresborough Rd	72 A3	Arklow Sq	73 B2
Miers Court Rd	72 B3	Artillery Rd	73 A2
Milstead Rd	72 A1	Augusta Rd	73 B2
Minster Rd	72 A1	Avebury Av	73 B1
Monmouth Clo	72 A1	Avenue Rd	73 B2
Mossy Glade	72 B3	Ayton Rd	73 A3
Napwood Clo	72 A3	Balmoral Pl	73 B2
Newnham Clo	72 A1	Bay View Rd	73 B1
Nightingale Clo	72 B3	Beechcroft Gdns	73 B2
Norfolk Clo	72 A1	Belgrave Clo	73 A2
Norreys Rd	72 B3	Bellevue Cotts	73 A2
Northumberland Av	72 B2	Bellevue Rd	73 B2
Nursery Rd	72 A2	Belmont Rd	73 A2
Oldfield Clo	72 A2	Belmont St	73 A2
Orchard St	72 B3	Beresford Rd	73 A3
Parkfield Rd	72 B1	Binnie Clo	73 B1
Patrixbourne Av	72 A1	Bolton St	73 A2
Pembury Wy	72 A1	Boughton Av	73 B1
Penshurst Clo	72 A1	Boundary Rd	73 A2
Petham Grn	72 A1	Brights Pl	73 A2
Pikefields	72 A1	Broad St	73 A2
Ploughmans Way	72 B3	Brockenhurst Rd	73 B2
Pluckley Clo	72 A1	Brunswick St	73 A2
Preston Way	72 A1	Camden Rd	73 A3
Pudding Rd	72 B2	Camden Sq	73 A2
Pump La	72 A1	Cannon Rd	73 A2

Cannonbury Rd	73 A3	Packers La	73 A2
Carlton Av	73 A3	Paradise	73 A2
Cavendish St	73 A2	Paragon St	73 A3
Cecilia Rd	73 A2	Paragon	73 A3
Central Rd	73 A2	Park Av	73 A1
Chapel Place La	73 A2	Park Chase	73 A1
Charles Rd	73 A2	Park Gate	73 A1
Chatham Pl	73 A2	Park Rd	73 A2
Chatham St	73 A2	Parkwood Clo	73 A1
Chester Av	73 A3	Penshurst Rd	73 B2
Church Hill	73 A2	Percy Rd	73 A2
Church Rd	73 A2	Plains of Waterloo	73 A2
Cliff St	73 A3	Poplar Rd	73 A2
Clifton Lawn	73 A3	Prestedge Av	73 A1
Codrington Rd	73 A3	Princes Rd	73 A2
Colburn Rd	73 B1	Priory Rd	73 A3
College Rd	73 A2	Promenade	73 B1
Cornhill	73 A3	Prospect Ter	73 A3
Cornwall Av	73 B2	Queen St	73 A3
Coronation Rd	73 A3	Queens Rd	73 B2
Cottage Rd	73 A3	Ramsgate Rd	73 B1
Cranbourne Clo	73 B2	Richmond Rd	73 A3
Crescent Rd	73 A2	Rodney St	73 A3
Cumberland Rd	73 A3	Rose Hill	73 A3
D'este Rd	73 B2	Rosebery Av	73 B1
Dane Cres	73 A2	Rosemary Av	73 B1
Dane Park Rd	73 A2	Rosemary Gdns	73 B1
Dane Rd	73 A2	Royal Cres	73 A3
Darren Rd	73 B1	Royal Parade	73 A3
Denmark Rd	73 A2	Royal Rd	73 A3
Dumpton Gap Rd	73 B1	Salisbury Av, Broadstairs	73 B1
Dumpton La	73 A1	Salisbury Av, Ramsgate	73 A2
Dumpton Park Dri	73 B2	Sanctuary Clo	73 B1
Dumpton Park Rd	73 A2	Sandwood Rd	73 B1
Duncan Rd	73 A3	School La	73 A2
Eagle Hill	73 A2	Seacroft Rd	73 B1
Eastern Rd	73 A2	Seven Stones Dri	73 B1
Effingham St	73 A3	Shaftesbury St	73 B2
Elham Way	73 B1	Shah Pl	73 A2
Elizabeth Rd	73 B3	Sherwood Gdns	73 A1
Ellen Av	73 A1	Shirley Av	73 A1
Ellington Rd	73 A2	Sion Hill	73 A3
Elms Av	73 A2	South Cliff Par	73 B1
Elmstone Rd	73 A2	Spencer Sq	73 A3
Ethelbert Rd	73 A3	Springfield Clo	73 A1
Fairfield Rd	73 A1	Staffordshire St	73 A2
Finsbury Rd	73 A2	Stanley Pl	73 A2
Flora Rd	73 A2	Stanley Rd	73 A2
Francis Gdns	73 B2	Staplehurst Av	73 B1
Freda Clo	73 B1	Station Approach Rd	73 A1
George Gdns	73 A3	Stonar Clo	73 A1
George St	73 A2	Sundew Gro	73 A2
George St	73 A3	Sussex St	73 A2
Gilbert Rd	73 A2	Sydney Rd	73 B2
Gordon Rd	73 A2	Tavistock Rd	73 A1
Grange Rd	73 A3	Thanet Rd	73 B2
Grange Way	73 B1	The Cloisters	73 A3
Grove Rd	73 A3	Thorn Gdns	73 A1
Grundys Hill	73 A3	Tomsons Pass	73 A2
Guildford Lawn	73 A3	Townley St	73 A3
Harbour Par	73 A3	Trinity Pl	73 B2
Harbour St	73 A3	Truro Rd	73 B2
Hardres Rd	73 A3	Turner St	73 A2
Hardres St	73 A2	Union Pl	73 A2
Harrison Rd	73 A2	Union Rd	73 B2
Hatfield Rd	73 A2	Unity Pl	73 B2
Hawkhurst Way	73 B1	Upper Dumpton Park Rd	73 A2
Hawthorn Clo	73 A1	Vale Rd	73 A3
Heathwood Dri	73 A1	Vale Sq	73 A3
Hereson Rd	73 A2	Vereth Rd	73 A3
Hertford Pl	73 A2	Victoria Par	73 B2
Hertford St	73 A3	Victoria Rd	73 B2
Hibernia St	73 A2	Vine Clo	73 A1
High St	73 A2	Waldron Rd	73 B1
Hillbrow Rd	73 A2	Wallwood Rd	73 B2
Hollicondane Rd	73 A2	Warten Rd	73 B1
Holly Rd	73 A2	Weatherly Dri	73 B1
Honeysuckle Rd	73 B2	Wellington Cres	73 B3
Ivy La	73 A3	West Cliff Prom	73 A3
James St	73 A3	West Cliff Rd	73 A3
Kent Pl	73 B3	West Dumpton La	73 A1
King St	73 A3	Western Esp	73 B1
Kings Rd	73 A2	Wickham Av	73 B2
Lenham Rd	73 B1	Willsons Rd	73 A3
Leonards Av	73 A1	Winstanley Cres	73 B2
Leopold Rd	73 A2	Winterstoke Cres	73 B2
Leopold St	73 A3	Winterstoke Undercliff	73 B2
Lillian Rd	73 B2	Winterstoke Way	73 B2
Liverpool Lawn	73 A3	Woodford Av	73 A1
Lyndhurst Rd	73 B2	York St	73 A3
Madeira Walk	73 A3	York Ter	73 A3
Margate Rd	73 A1		
Marina Esplanade	73 B3		
Marina Rd	73 B2	**ROCHESTER**	
Marlborough Rd	73 A3	Abbots Clo	74 A3
Meeting St	73 A2	Albany Rd	74 B2
Michael Av	73 B2	Albert Rd	74 B2
Military Rd	73 A3	Alma Pl	74 A1
Mill Cotts	73 A3	Almon La	74 B2
Minster Clo	73 B1		
Montague Rd	73 A2		
Montefiore Av	73 B1		
Muir Rd	73 B1		
Mulberry Clo	73 B2		
Nelson Cres	73 A3		
Newcastle Hill	73 A2		
Newlands Rd	73 A1		
Nicholls Av	73 B1		
North Av	73 A3		
Ocean Vw	73 B1		

High St	78 B2
Hogs Corner	78 B2
Honfleur Rd	78 A2
Hythe Pl	78 B2
Johns Green	78 A3
Jubilee Rd	78 A2
King St	78 B2
Knightrider St	78 B2
Laburnam Av	78 A2
Loop St	78 A2
Loop St Mews	78 A1
Love La	78 B2
Manwood Rd	78 B2
Mill Clo	78 B2
Millwall Pl	78 B2
Moat Sole	78 A2
New Romney Pl	78 A2
New St	78 A2
Paradise Row	78 A1
Pondicherry All	78 B2
Potter St	78 B2
Poulders Gdns	78 A2
Poulders Rd	78 A2
Quay La	78 B2
Ramsgate Rd	78 B1
Richborough Rd	78 A1
St Andrews Lees	78 B2
St Barts Rd	78 A2
St Georges Lees	78 B2
St Georges Rd	78 B2
St Peters St	78 B2
Sandown Rd	78 B2
Sandwood Rd	78 A2
Sarre Pl	78 A2
School Rd	78 A1
Short La	78 B2
Stonar Clo	78 B1
Stonar Gdns	78 B1
Stone Cross Lees	78 A3
Strand St	78 A1
Sunnyside Gnds	78 A2
The Butchery	78 B1
The Causeway	78 A1
The Chain	78 B2
The Crescent	78 A3
The Quay	78 B2
Upper Strand St	78 B2
Vicarage La	78 A1
Wantsume Lees	78 A1
Whitefriars Meadow	78 A2
Whitefriars Way	78 A2
Woodnesborough Rd	78 A3

SEAL

Ash Platt Rd	79 A2
Bentleys Meadow	79 A1
Childsbridge La	79 A1
Childsbridge Way	79 A1
Church Field Cotts	79 A1
Church Rd	79 A2
Church St	79 A2
Copse Bank	79 A1
Grove Rd	79 A2
High St	79 A2
Highlands Park	79 A2
Honey Pot La	79 B1
Jubilee Rise	79 A2
Knox Ct	79 A1
Landway	79 A1
Maidstone Rd	79 A1
Meadowlands	79 A1
Mills Cres	79 A1
Park La	79 A2
Pudding La	79 A2
Ragge Way	79 A1
School La	79 A2
Seal Dri	79 A2
Seal Rd	79 A2
Sealcroft Cotts	79 A1
The Sheilings	79 A1
Wildernesse Rd	79 A2
Wilmar Way	79 A1
Zambra Way	79 A1
Zion St	79 A2

SEVENOAKS

Akehurst La	80 B3
Amherst Rd	80 A2
Argyle Rd	80 A3
Ashley Clo	80 A2
Ashley Rd	80 A2
Avenue Rd	80 A2
Bank St	80 A3
Bayham Rd	80 B2
Beech Rd	80 A3
Belmont Rd	80 A2
Berwick Way	80 A1
Bethel Rd	80 B2
Birch Clo	80 A2
Blackhall La	80 B2
Blair Dri	80 A2
Blighs Rd	80 A3
Bosville Av	80 A2
Bosville Dri	80 A2
Bosville Rd	80 A2
Bouchier Clo	80 A3
Bradbourne Park Rd	80 A2
Bradbourne Rd	80 A1
Bradbourne Vale Rd	80 A1
Brewery La	80 A3
Buckhurst Av	80 A3
Buckhurst La	80 A3
Camden Rd	80 A1
Carrick Dri	80 A2
Cavendish Av	80 A1
Cedar Ter	80 B2
Chancellor Way	80 A1
Charter House Dri	80 A2
Chartway	80 B3
Chatham Hill Rd	80 A1
Chestnut La	80 A2
Clarendon Rd	80 A3
Clockhouse La	80 A2
Cobden Rd	80 B2
Coombe Av	80 A1
Cramptons Rd	80 A1
Crownfields	80 A3
Dartford Rd	80 A3
Dorset St	80 A3
Eardley Rd	80 A3
Egdean Walk	80 A2
Farm Rd	80 B1
Garden Rd	80 B1
Garyock Dri	80 A3
Golding Rd	80 A1
Gordon Rd	80 A3
Granville Rd	80 A2
Greatness La	80 B1
Greatness Rd	80 B1
Grove Rd	80 B1
Harrison Way	80 A1
Hartslands Rd	80 B2
High St	80 A3
Hill Crest	80 A2
Hillborough Av	80 A3
Hillingdon Av	80 B1
Hillingdon Rise	80 B1
Hillside Rd	80 B2
Hitchen Hatch La	80 A2
Holly Bush Clo	80 B2
Holly Bush Ct	80 B2
Holly Bush La	80 B2
Hollyoake Ter	80 A2
Holmesdale Rd	80 B2
Hospital Rd	80 B1
Hunsdon Dri	80 A2
Kennedy Gdns	80 B2
Kincraig Dri	80 A2
Kippington Rd	80 A3
Kirk Ct	80 A2
Knole La	80 B3
Knole Rd	80 B2
Knole Way	80 B3
Knotts Pl	80 A2
Lambarde Dri	80 A2
Lambarde Rd	80 A1
Lime Tree Walk	80 A3
Linden Chase Rd	80 A2
Little Wood	80 B1
Locks Yard	80 A3
London Rd	80 A3
Lyle Park	80 A2
Market Pl	80 A3
Meadow Clo	80 A2
Merlewood	80 A2
Mill La	80 B1
Mount Harry Rd	80 A2
Nicholson Way	80 B1
Northview Rd	80 B1
Nursery Rd	80 B2
Oak Hill	80 A2
Oak Hill Rd	80 A3
Oak La	80 A3
Oak Sq	80 A3
Oakdene Rd	80 A2
Oakfields	80 A3
Orchard Clo	80 B1
Otford Rd	80 A1
Park La	80 A2
Pembroke Rd	80 A3
Pendennis Rd	80 A2
Pine Needle La	80 A2
Pinehurst	80 B1
Pinewood Av	80 B1
Plymouth Dri	80 A3
Plymouth Park	80 B3
Pound La	80 A2
Prospect Rd	80 B2
Quaker Clo	80 B2
Quakers Hall La	80 B2
Quarry Hill	80 B2
Queens Dri	80 B1
Rectory La	80 A3
Rockdale Rd	80 A3
Rose Field	80 A2
Sackville Clo	80 A1
St Botolphs Av	80 A2
St Botolphs Rd	80 A2
St Georges Rd	80 A1
St James Rd	80 A1
St Johns Hill	80 A1
St Johns Rd	80 A1
Sandy La	80 B2
Seal Hollow Rd	80 B2
Seal Rd	80 B1
Serpentine Rd	80 B2
Six Bells La	80 A3
South Park	80 A3
Station Par	80 A2
Suffolk Way	80 A3
Swaffield Rd	80 B1
The Crescent	80 B1
The Dene	80 A3
The Drive	80 A2
The Glade	80 A2
The Green	80 B1
The Shambles	80 A3
Thicketts	80 A2
Tubs Hill	80 A2
Tubs Hill Par	80 A2
Vale Rd	80 A1
Valley Rd	80 A3
Victoria Rd	80 A3
View Rd	80 A2
Vine Av	80 A2
Vine Court Rd	80 A2
Warren Ct	80 A3
Weavers La	80 B1
Webbs Alley	80 B3
Well Ct	80 A3
Westfield	80 B1
Wickenden Rd	80 B1
Wilderness Av	80 B1
Wilderness Mt	80 B1
Winchester Gro	80 A2
Wood Lodge Grange	80 A1
Woodland Rise	80 B2
Woodside Rd	80 A2
Yeomans Meadow	80 A3

SHEERNESS

Acorn St	81 B2
Albion Pl	81 B2
Alden Clo	81 A3
Alma Rd	81 B2
Almond Tree Clo	81 A3
Anchor La	81 A1
Appledore Av	81 B3
Archway Rd	81 A1
Beach St	81 B1
Berridge Rd	81 B2
Boxley Clo	81 B3
Bredhurst Clo	81 B3
Bridge St	81 B1
Brielle Way	81 A2
Broad St	81 B2
Broadway	81 B1
Carlton Av	81 B2
Cavour Rd	81 B2
Cecil Av	81 B2
Chapel St	81 A1
Charles St	81 A1
Cherry Tree Clo	81 A3
Chilham Clo	81 B3
Clarence Row	81 B1
Coats Av	81 A3
Cromwell Rd	81 A3
Cross St	81 B1
Davie Clo	81 B3
Delamark Rd	81 B1
Detling Clo	81 B3
Dock Rd	81 A1
Dorset Rd	81 A2
East La	81 A1
Edenbridge Dri	81 B3
Esplanade	81 B1
Estuary Rd	81 B2
First Av	81 B2
Fleet Av	81 B2
Galway Rd	81 B2
Grace Rd	81 A2
Granville Pl	81 B2
Granville Rd	81 B2
Hare St	81 B1
Harris Rd	81 B2
Hartlip Clo	81 B3
Hawthorn Av	81 A3
High St	81 A1
Holland Clo	81 B2
Hope St	81 B2
Invicta Rd	81 B2
Kent Rd	81 B2
Kent St	81 A1
King St	81 A1
Kings Head Alley	81 A1
Larch Ter	81 A3
Linden Dri	81 A3
Main Rd	81 A1
Maple St	81 B2
Medway Rd	81 B2
Meyrick Rd	81 B1
Milstead Clo	81 B3
Nelson Clo	81 A2
New Rd	81 A2
New St	81 B2
Newcomen Rd	81 B1
Newland Rd	81 A3
Pepys Av	81 B1
Pier Rd	81 A3
Portland Ter	81 B2
Queens Way	81 A3
Railway Rd	81 B2
Ranelagh Rd	81 B1
Regents Pl	81 B1
Rose St	81 B2
Royal Rd	81 B1
Rule Ct	81 B2
Russell St	81 B2
St Agnes Gdns	81 B2
St Georges Av	81 B3
St Georges Ct	81 B2
School La	81 A1
Second Av	81 B2
Sheppey St	81 A1
Short St	81 B1
Shrubsole Av	81 B2
South View Gdns	81 B2
Strode Cres	81 B1
Swale Av	81 B2
Thames Av	81 B2
Trinity Pl	81 B1
Trinity Rd	81 B1
Union St	81 A1
Victoria St	81 B2
Victory St	81 B1
Vincent Gdns	81 B2
West La	81 A1
West St	81 A1
Wheatsheaf Gdns	81 B2
Whiteway Rd	81 A3
Winstanley Rd	81 B1
Wood St	81 B2

SHORNE

Bowsden Rd	79 A4
Brewers Rd	79 A3
Burdett Av	79 B4
Butchers Hill	79 A3
Cobb Dri	79 B3
Court Lodge	79 A3
Coutts Av	79 B4
Crown Grn	79 B3
Crown La	79 B3
Forge La	79 B3
Gravesend Rd	79 B4
Green Farm La	79 B4
Holland Clo	79 A3
Malthouse La	79 B3
Manor Field	79 B3
Mill Hill La	79 B3
Peartree La	79 A4
Pondfield La	79 A4
Racefield Clo	79 A3
Racefield Clo	79 A3
Swillers La	79 A3
Tanyard Hill	79 A4
The Ridgeway	79 A3
The Street	79 B3
Warren Vw	79 A4
Woodlands La	79 A3

SITTINGBOURNE

Addington Rd	82 A2
Adelaide Dri	82 A2
Albany Rd	82 A2
Arthur St	82 A1
Arundel Av	82 B3
Aubretia Walk	82 B2
Avenue of Remembrance	82 A2
Balmoral Ter	82 A2
Barrow Gro	82 A2
Bassett Rd	82 A2
Beechwood Av	82 A1
Bell Rd	82 B2
Belmont Rd	82 A2
Berkeley Ct	82 A2
Berry St	82 B2
Blandford Gdns	82 A3
Borden La	82 A2
Bourne Gro	82 A1
Bradley Dri	82 A3
Brenchley Rd	82 B3
Brewery Rd	82 B1
Burley Rd	82 A2

Capel Rd	82 A2
Central Av	82 B2
Chalkwell Rd	82 A1
Chappell Way	82 A1
Charlotte St	82 A1
Chartwell Gro	82 A2
Chaucer Rd	82 A2
Chegworth Gdns	82 A3
Cherry Clo	82 A1
Chilton Av	82 B2
Church St	82 A1
Clarendon Clo	82 B3
Cobham Av	82 A3
College Rd	82 A2
Connaught Rd	82 A2
Cooks La	82 B1
Courtland Clo	82 A1
Courtland Mews	82 A1
Cranbrook Dri	82 A3
Crescent St	82 B2
Cross La	82 A1
Crossways	82 B3
Crown Quay La	82 B1
Crown Rd	82 A1
Dean Rd	82 A1
Derby Clo	82 A1
Dobbie Clo	82 A1
Does All	82 B2
Dover St	82 A2
Downs Clo	82 A3
East St	82 B2
Eastwood Rd	82 A1
Epps Rd	82 A2
Eurolink Way	82 B2
Fairview Rd	82 B2
Falcon Ct	82 B3
Farm Cres	82 B3
Forge Rd	82 A1
Fountain Ct	82 A2
Frederick St	82 A2
Fulston Pl	82 B2
Gas Rd	82 B1
Gerrards Dri	82 B3
Gibson St	82 A2
Glovers Cres	82 B2
Gore Court Rd	82 A3
Grafton Rd	82 B2
Grafton Way	82 B2
Grayshott Clo	82 B2
Hales Rd	82 A3
Hall Clo	82 B1
Hanover Clo	82 A2
Harrier Dri	82 B2
Harvey Dri	82 B3
Hawthorn Rd	82 A1
Haysel	82 B3
Heather Clo	82 B2
High St, Milton Regis	82 A1
High St, Sittingbourne	82 B2
Highstead Rd	82 B2
Hill Brow	82 A2
Hobart Gdns	82 A2
Homewood Av	82 A2
Howard Av	82 A1
Hythe Rd	82 A1
Johnson Rd	82 A2
Jubilee St	82 A1
Kent Av	82 A2
Kestrel Clo	82 B2
Kiln Clo	82 B2
King St	82 A1
Kingsmill Clo	82 A1
Laburnum Pl	82 A1
Lammas Dri	82 A1
Larkfield Av	82 A1
Lavender Ct	82 B2
Laxton Way	82 A1
Lime Gro	82 B2
Linden Clo	82 A2
Little Glovers	82 B2
London Rd	82 A1
London Road Trading Est	82 A2
Lydbrook Clo	82 A2
Lyndhurst Gro	82 A3
Manor Gro	82 A2
Manwood Clo	82 B3
Medway Clo	82 A2
Merlin Clo	82 B2
Mill Ct	82 B2
Mill La	82 B1
Mill Way	82 A1
Millen Rd	82 A1
Millfield	82 B2
Milton Rd	82 B1
Minterne Av	82 A3
Musgrave Rd	82 B1
Northwood Dri	82 B3
Orchard Pl	82 B2
Oyster Clo	82 B1
Park Av	82 A3

Park Dri	82 A3
Park Rd	82 A2
Pembury St	82 B2
Peregrine Dri	82 B2
Periwinkle Clo	82 A1
Pond Dri	82 B2
Princes St	82 B1
Railway Ter	82 B2
Regency Ct	82 A1
Riverhead Clo	82 A2
Rock Rd	82 A2
Roman Sq	82 B2
Romney Ct	82 A1
Roonagh Ct	82 A3
Roseleigh Rd	82 A3
St Michaels Clo	82 B2
St Michaels Rd	82 B2
St Pauls Rd	82 A1
Shakespeare Rd	82 B2
Shortlands Rd	82 B2
Shrubland Av	82 B3
Silverdale Gro	82 A2
South Av	82 B2
Springfield Rd	82 A1
Stanhope Av	82 B2
Staple Clo	82 A1
Staplehurst Rd	82 A1
Station Pl	82 B2
Station St	82 B2
Sterling Rd	82 A3
Sydney Av	82 A2
Tavistock Clo	82 A2
The Burrs	82 B2
The Butts	82 B2
The Fairway	82 A3
The Fieldings	82 A3
The Finches	82 B2
The Forum	82 B2
The Meadow	82 B3
The Roundel	82 B3
The Wall	82 B1
Trotts Hall Gdns	82 B2
Tunstall Rd	82 A3
Unity St	82 A2
Upton La	82 A2
Valenciennes Rd	82 A2
Victoria Rd	82 A1
Viners Clo	82 B3
Walmer Gdns	82 A1
Waterloo Rd	82 A1
Watling Pl	82 B2
Watsons Hill	82 A1
Weald Ct	82 A3
Well Winch Rd	82 A1
West La	82 B2
West Ridge	82 A2
West St	82 A2
Westbourne St	82 A1
Westerham Rd	82 A2
Wharf Way	82 B1
William St	82 A2
Windermere Gro	82 A2
Windmill Rd	82 A1
Windsor Dri	82 A3
Woodcourt Clo	82 A3
Woodside Gdns	82 A3
Woodstock Rd	82 A3
Worcester Dri	82 A1

SNODLAND

Annie Rd	83 A3
Apple Clo	83 B3
Ashbee Clo	83 B2
Bingley Clo	83 A2
Birling Rd	83 A2
Bramley Rd	83 B2
Brook La	83 A3
Brook St	83 B2
Bull Fields	83 B2
Busbridge Rd	83 A2
Cemetery Rd	83 A1
Chapel Rd	83 B2
Charles Clo	83 B2
Church Field	83 B1
Constitution Hill	83 A2
Cooper Rd	83 A3
Covey Hall Rd	83 B2
Coxs Clo	83 A2
Dowling Clo	83 A2
Dryland Rd	83 A2
East St	83 B2
Freelands Rd	83 A2
Gassons Rd	83 A2
Godden Rd	83 A2
Gorham Clo	83 A2
High St	83 B2
Hodgson Cres	83 B2
Holborough Rd	83 B2
Hollow La	83 A2
Hook Rd	83 A2
Kent Rd	83 B3

Ladds La	83 A1
Lakeside Vw	83 A3
Lakeview Clo	83 B3
Lee Rd	83 B2
Lucus Rd	83 A2
Malling Rd	83 A3
May St	83 B2
Medway Walk	83 A2
Midsummer Rd	83 A2
Morhen Clo	83 A2
Nevill Pl	83 B2
Nevill Rd	83 B3
Norman Rd	83 B3
Orchard Way	83 A2
Oxford St	83 B2
Paddlesworth Rd	83 A2
Portland Pl	83 B2
Pout Rd	83 A2
Pridmore Rd	83 A2
Queens Av	83 B2
Queens Rd	83 B2
Recreation Av	83 B2
Rectory Clo	83 B2
Ritch Rd	83 A2
Roberts Rd	83 A2
Rocfort Rd	83 B2
Roman Rd	83 A2
St Benedict Rd	83 A2
St Katherines La	83 A2
Saltings Rd	83 B3
Sandy La	83 A3
Sharnel La	83 B2
Simpson Rd	83 B3
Snodland By-Pass	83 A3
Snodland Rd	83 A3
Taylor Rd	83 A2
The Bulrushes	83 B2
The Grooves	83 A2
Thomson Clo	83 B2
Tomlin Clo	83 B2
Townsend Rd	83 A2
Vauxhall Cres	83 A3
Veles Rd	83 A2
Waghorn Rd	83 B2
Whitedyke Rd	83 A1
Woodlands Av	83 A2
Wyvern Clo	83 A2

SOUTHBOROUGH

All Saints Rd	84 B3
Andrew Rd	84 B2
Argyle Rd	84 B1
Barnetts Way	84 B2
Bedford Rd	84 A2
Beltring Rd	84 A3
Birchwood Av	84 A1
Bounds Oak Way	84 A1
Breedon Av	84 A2
Brian Cres	84 B2
Brightridge	84 A2
Brokes Way	84 B2
Brookhurst Gdns	84 A1
Broomhill Park Rd	84 A2
Cambrian Rd	84 B3
Carville Av	84 A2
Castle Rd	84 A1
Charles St	84 A2
Chestnut Av	84 B2
Chestnut Clo	84 B2
Church Rd	84 A1
Colebrook Rd	84 B3
Colonels Way	84 B1
Constitution Hill Rd	84 A2
Crendon Park	84 A2
Cunningham Rd	84 B3
Darnley Dri	84 A1
Denbigh Rd	84 B3
Doone Brae	84 B1
Doric Av	84 A2
Doric Clo	84 A2
Dower House Cres	84 A1
Draper St	84 A1
Dynevor Rd	84 B3
East Cliff Rd	84 B3
Edward St	84 A2
Elm Rd	84 A2
Fairlight Clo	84 B1
Fernhurst Gdns	84 B1
Forge Rd	84 A2
Garlinge Rd	84 A2
Gordon Rd	84 B3
Great Bounds Dri	84 A1
Great Brooms	84 B2
Grosvenor Rd Nth	84 B3
Harlands Way	84 A1
High Brooms Rd	84 B2
Highfield	84 B3
Hill Crest	84 B2
Hill Garth	84 B3
Holden Corner	84 A2
Holden Park Rd	84 A2

Holden Rd	84 A2
Holmewood Rd	84 B3
Hopwood Gdns	84 B3
Horizon Clo	84 B2
Hythe Clo	84 A2
Impala Gdns	84 B3
Keel Gdns	84 A2
Kibbles La	84 A2
Ladys Gift Rd	84 A2
Laurel Bank	84 B3
Leighton Clo	84 A3
Littlebounds Clo	84 A1
London Rd	84 A1
Manor Rd	84 A2
Meadow Rd	84 A2
Mereworth Rd	84 B3
Merrion Clo	84 B3
Montgomery Rd	84 B3
New England Rd	84 B3
Newlands Rise	84 B3
Newlands Road	84 B3
Newlands Way	84 B3
Norstead Gdns	84 B3
Norton Rd	84 A2
Nursery Rd	84 B3
Oak End Clo	84 B2
Park House Gdns	84 A2
Park Rd	84 A2
Pennington Rd	84 A1
Pinewood Gdns	84 A2
Powdermill Clo	84 B2
Powdermill La	84 B3
Prospect Pk	84 A2
Prospect Rd	84 A2
Reynolds La	84 A3
Riddlesdale Av	84 B3
Ruscombe Clo	84 A2
St Andrews Park Rd	84 A2
St Davids Rd	84 B3
St Johns Pk	84 A2
St Johns Rd	84 B3
St Lukes Rd	84 B3
St Michaels Rd	84 B3
Salisbury Rd	84 B2
Sheffield Rd	84 A2
Silverdale La	84 B3
Silverdale Rd	84 B3
Sir Davids Park	84 A2
Smythe Clo	84 A1
South View Rd	84 B3
Southfield Rd	84 A3
Southfields Way	84 B3
Southwood Av	84 B3
Speldhurst Rd	84 A2
Springfield Rd	84 A2
Stephens Rd	84 B3
Stewart Rd	84 B3
Sumhill Av	84 A2
Taylor St	84 A2
Tedder Rd	84 B3
The Crescent	84 A1
The Fairways	84 B3
The Ridgeway	84 B2
Upper Dunstan Rd	84 B3
Vale Av	84 A2
Vale Rd	84 A2
Valley Vw	84 B2
Vauxhall La	84 A1
Vicarage Rd	84 A1
Victoria Rd	84 A2
Weare Rd	84 B2
Welbeck Av	84 B3
West Park Av	84 A2
Western Rd	84 A2
Wilman Rd	84 B3
Wolsey Rd	84 B3
Woodlands Rd	84 B3
Wooley Clo	84 A2
Wooley Rd	84 A2
Yew Tree Rd	84 A2

STAPLEHURST

Allen Sq	85 B2
Bathurst Clo	85 A2
Bathurst Rd	85 A2
Bell La	85 A2
Benden Clo	85 B2
Bower Walk	85 A2
Brooms Clo	85 A1
Butcher Rd	85 A2
Chapel La	85 B2
Chestnut Av	85 A2
Church Grn	85 A2
Clapper La	85 A1
Corner Farm Rd	85 A1
Cornforth Clo	85 B2
Crowther Clo	85 A2
Fir Tree Clo	85 B2
Fishers Clo	85 B1
Fishers Rd	85 B1
Fletcher Rd	85 A2

Frittenden Rd	85 B3			

STONE

STROOD

STURRY

Chestnut Dri	88 B1		
Church La	88 A2		
Copt Clo	88 B1		
Deansway Av	88 A1		
Delaware Clo	88 B2		
Denne Clo	88 B1		
Fairview Gdns	88 B2		
Field Way	88 B2		
Fordwich Rd	88 B2		
Forge Clo	88 B2		
Hawe La	88 B1		
Heath Clo	88 A1		
Herne Bay Rd	88 A1		
High St, Fordwich	88 B3		
High St, Sturry	88 B2		
Hillbrow Av	88 A1		
Hoades Wood Rd	88 B1		
Homewood Rd	88 B2		
Hudson Clo	88 B1		
Ince Rd	88 B1		
Island Rd	88 B2		
Laburnham La	88 B1		
Ladywood Rd	88 B1		
McCarthy Av	88 A1		
Mayton Rd	88 A1		
Meadow Rd	88 B2		
Mill Rd	88 A3		
Milner La	88 A2		
Oakwood Rd	88 B1		
Park Vw	88 A2		
Pleydell Cres	88 B1		
Popes La	88 A1		
Redcott La	88 B1		
Risdon Clo	88 B2		
Riverview	88 A1		
Rowan Clo	88 B2		
St Nicholas Clo	88 B1		
Shalloak Rd	88 A2		
Sleigh Rd	88 B2		
Spring La	88 B3		
Staines Hill	88 B1		
Sturry Hill	88 A2		
Sturry Rd	88 A3		
Sweechgate	88 A1		
Tennyson Rd	88 A3		
The Drove	88 B3		
Twyne Clo	88 B2		
Well Clo	88 B2		
Whatmer Clo	88 B2		
Woodside	88 B1		

SWANLEY

Abbotts Clo	90 B2	Greenacre	90 A2
Acacia Walk	90 A1	Greenside	90 A1
Alder Way	90 A1	Hart Dyke Cres	90 A2
Alexandra Clo	90 B1	Hart Dyke Rd	90 A1
Almond Dri	90 A1	Harvest Way	90 A2
Archer Way	90 B1	Haven Clo	90 B1
Ash Clo	90 A1	Hazel End	90 B3
Aspen Clo	90 A1	Heather End	90 A3
Azalea Dri	90 A2	Heathfield Ter	90 A1
Bartholomew Way	90 B2	Heathwood Gdns	90 A1
Beech Av	90 B2	Hewitt Pl	90 A2
Bevan Pl	90 B2	High Firs	90 A2
Birchwood Park Av	90 B2	High St	90 B2
Birchwood Rd	90 A1	Highlands Hill	90 B1
Bonney Way	90 B1	Hilda May Av	90 A1
Bourne Way	90 A2	Hollytree Av	90 A1
Bramley Clo	90 A2	Homefield Clo	90 A2
Bremner Clo	90 A2	Irving Way	90 A1
Broadway	90 A3	Juniper Walk	90 A1
Brook Rd	90 A2	Kettlewell Ct	90 B1
Cedar Clo	90 A1	Kingswood Av	90 B2
Chapel Rd	90 A3	Laburnum Av	90 A2
Charnock	90 B2	Ladds Way	90 A2
Cherry Av	90 A2	Larch Walk	90 A1
Church Rd	90 A3	Lavender Hill	90 A3
College Rd	90 A1	Lawn Clo	90 A1
Conifer Way	90 A1	Leechcroft Av	90 B2
Cornel Gdns	90 A1	Leewood Pl	90 A2
Court Cres	90 B2	Lesley Clo	90 A2
Cranleigh Dri	90 B2	Leydenhatch La	90 A1
Cray Rd	90 A3	Leyhill Clo	90 A2
Crescent Gdns	90 A1	Lila Pl	90 A2
Cyclamen Rd	90 A2	Lilac Gdns	90 A2
Dahlia Dri	90 B1	Lime Rd	90 A2
Dale Clo	90 A1	London Rd	90 A1
Dale Rd	90 A1	Lower Croft	90 B2
Downsview Clo	90 B1	Lulling Stone Av	90 B2
Edgar Clo	90 B2	Lynden Way	90 A1
Edward Gdns	90 A2	Main Rd, Crockenhill	90 A3
Egerton Av	90 B1	Main Rd, Hextable	90 B1
Elm Dri	90 A1	Manse Way	90 B2
Everest Pl	90 A2	Maple Clo	90 B1
Eynsford Rd	90 A3	Mark Way	90 B3
Farm Av	90 A2	Mayes Clo	90 B2
Five Wents	90 B1	Mead Clo	90 B3
Glendale	90 B3	Millbrow	90 B1
Goldsel Rd	90 A3	Montague Rd	90 B1
Green Court Rd	90 A3	Moreton Clo	90 B1
		Moultain Hill	90 A1
		New Barn Rd	90 A1
		New Rd	90 B2
		Newlands Est	90 B1
		Nightingale Way	90 A2
		Northview	90 A1
		Nursery Clo	90 A1
		Nutley Clo	90 B1
		Old Farm Gdns	90 B2
		Oliver Rd	90 A2
		Over Mead	90 B3
		Park Rd	90 A1
		Pear Tree Clo	90 A1
		Pemberton Gdns	90 B2
		Phillip Av	90 A2
		Pine Clo	90 B2
		Pinks Hill	90 A2
		Reeves Cres	90 A2
		Rogers Ct	90 B2
		Rowan Rd	90 A2
		Russet Way	90 A1
		Ruxton Clo	90 A2
		St Georges Rd	90 B2
		St Lukes Clo	90 A1
		St Marys Rd	90 A1
		Salisbury Rd	90 B2
		Selah Dri	90 A1
		Sermon Dri	90 A2
		Seven Acres	90 A3
		Sheridan Clo	90 B2
		Shurlock Av	90 A1
		Sounds Lodge	90 A3
		Southern Pl	90 A2
		Southview Clo	90 B2
		Springfield Av	90 B2
		Station App	90 A2
		Station Rd	90 B2
		Stone Cross Rd	90 A1
		Strawberry Field	90 A1
		Stuart Clo	90 B1
		Swanley By Pass	90 A2
		Swanley La	90 B2
		Sycamore Dri	90 A2
		The Birches	90 B1
		The Croft	90 A2
		The Green	90 A3
		The Oaks	90 A1
		The Orchard	90 A1
		Victoria Hill Rd	90 B1
		Walnut Way	90 A1
		Wansbury Way	90 B3
		Waylands	90 B2
		West Harold	90 A2

West View Rd	90 B2
West View Rd, Crockenhill	90 A3
Wested La	90 B3
White Crofts	90 A1
Willow Av	90 B2
Wisteria Gdns	90 A1
Woodgers Gro	90 B1
Woodlands Clo	90 B2
Woodlands Rise	90 B1
Woodview Rd	90 A1

TEMPLE EWELL

Alkham Valley Rd	91 A2
Brook Side	91 A1
Church Hill	91 A1
Courtlands Dri	91 B2
Down Hill Clo	91 B1
Egerton Rd	91 B1
Green La	91 B1
High St	91 B1
Kearnsey Av	91 B2
Laburnham Clo	91 B2
London Rd	91 A1
Lower Rd	91 B1
Malvern Meadow	91 B2
Malvern Rd	91 B1
Mill St	91 B1
Park Rd	91 B1
Pavilion Meadow	91 B2
Redvers Cotts	91 B2
Target Firs	91 B1
Templar Rd	91 A1
Temple Clo	91 B1
Templeside	91 B1
The Avenue	91 B1
Watersend	91 A1
Wellington Rd	91 B1
Woodside Clo	91 B2

TENTERDEN/ ST MICHAELS

Adams Clo	92 A2
Admirals Walk	92 B2
Appledore Rd	92 B3
Ashford Rd	92 A1
Austins Orchard	92 A3
Barnfield	92 B1
Beacon Oak Rd	92 A2
Beacon Walk	92 B2
Bells Clo	92 A3
Bells La	92 A3
Bennett Mews	92 A3
Bridewell La	92 A3
Burgess Row	92 A3
Caxton Clo	92 A3
Chalk Av	92 A1
Cherry Orchard	92 A3
Chestnut Clo	92 B2
Church Rd	92 A3
Collison Pl	92 B3
Coombe La	92 A3
Craythorne	92 B2
Crisfield Ct	92 A2
Curteis Rd	92 A2
Danemore	92 A3
Drury Rd	92 A2
East Cross	92 A3
East Hill	92 A3
Eastgate	92 B2
Eastwell Barn Mews	92 A3
Eastwell Meadows	92 A3
Elmfield	92 A3
Forson Clo	92 A2
Glenwood Clo	92 A1
Golden Sq	92 A3
Grange Cres	92 A1
Grange Rd	92 A1
Haffenden Rd	92 A2
Hales Clo	92 A3
Heather Dri	92 A1
Henley Fields	92 A2
Henley Vw	92 A2
High St	92 A3
Highbury La	92 A3
Homewood Rd	92 A2
Ingleden Park Rd	92 B2
Kiln Field	92 B3
Knockwood Rd	92 B2
Leslie Cres	92 A2
Limes Clo	92 B3
Longfield	92 A3
Malt House La	92 A3
Marshalls Land	92 A1
Martins Clo	92 B2
Mayors Pl	92 A3
Mill La	92 A2
Mount Pleasant	92 B2
Oaks Rd	92 A3
Orchard Rd	92 A1

Ox La	92 B1
Park View	92 A3
Parkside St	92 A3
Penderel Mews	92 A3
Pittlesden	92 A3
Pope House La	92 A1
Priory Way	92 B3
Recreation Grnd Rd	92 A3
Rogersmead	92 A3
Rothley Clo	92 A2
St Benets Ct	92 A2
St Michaels Clo	92 A3
St Michaels Rd	92 A1
Sandy La	92 A3
Sayers Ct	92 A3
Sayers La	92 A3
Sayers Sq	92 A3
Shoreham La	92 A1
Shrubcote	92 B3
Silver Hill	92 A2
Six Fields Path	92 A3
Smallhythe Rd	92 A3
Southgate Rd	92 B3
Springfield Av	92 B2
Stace Clo	92 B3
Station Rd	92 A3
Summer Clo	92 B2
Swain Rd	92 A1
Tannery Clo	92 A2
The Lindens	92 A2
The Pavement	92 A1
Three Fields Path	92 A3
Turners Av	92 A3
Vineys Gdns	92 B2
Wayside Av	92 A1
Wayside Little Hill	92 A1
Wealden Av	92 B2
Wells Clo	92 A2
West Cross Gdns	92 A3
Westwell Ct	92 A3
Wights Clo	92 A3
Woodchurch Rd	92 B2

TESTON

Church St	91 A4
Courtlands	91 A4
Courtlands Clo	91 A4
Fairlawn Clo	91 A3
Livesey St	91 A3
Malling Rd	91 A3
Nestor Ct	91 A4
Red Hill	91 A3
Teston La	91 B4
The Street	91 A4
Tonbridge Rd	91 A4
Woodlands Clo	91 A4

TEYNHAM

Amber Clo	93 A1
Baker Clo	93 A1
Belle Friday Clo	93 A1
Bradfield Av	93 A1
Cellar Rd	93 A2
Cherry Gdns	93 A2
Cherry Tree Clo	93 A1
Conyer Rd	93 A1
Deerton St	93 B1
Donald Moor Av	93 A2
Frognal Clo	93 A2
Frognal La	93 A1
Harrys Rd	93 A1
Honey Ball Walk	93 A1
Lewson St	93 B2
London Rd	93 A2
Lower Rd	93 A1
Lynsted Way	93 A2
Morello Clo	93 A1
Noads La	93 A2
Nobel Clo	93 A2
Nutberry Clo	93 A1
Orchard Vw	93 A1
Rivers Rd	93 A1
Roper Rd	93 A1
Rundle Clo	93 A1
Sittingbourne Rd	93 A2
Station Rd	93 A2
The Cres	93 A1
Triggs Row	93 A1

TONBRIDGE

Albany Clo	94 B3
Albert Rd	94 A2
Alexandra Rd	94 A2
Angel La	94 B2
Angel Walk	94 B2
Ashburnham Rd	94 B1
Audley Av	94 A2
Avebury Av	94 A2

Stratford St	95 B1	Gladstone Rd	96 B1	Hanbury Clo	60 B3	Garlinge	98 B2
Stuart Clo	95 A3	Grams Rd	96 A3	Hillside Ct	60 B3	Dent-de-Lion Clo	98 B3
Surrey Clo	95 A3	Granville Rd	96 B2	Leney Dri	60 B4	Domneva Rd	98 A2
Sussex Clo	95 B3	Green La	96 A3	Lodge Clo	60 A3	Edinburgh Rd	98 A2
Sutherland Rd	95 B2	Greenacre Dri	96 B3	Love La	60 A3	Edith Rd	98 A2
The Chase	95 B2	Grove Rd	96 B1	Mill La	60 A3	Edmanson Av	98 A2
The Drive	95 B3	Guildford Ct	96 B2	Old Rd	60 A3	Egbert Rd	98 A2
The Ferns	95 B1	Halstatt Rd	96 A2	Phoenix Dri	60 B4	Ellington Av	98 B3
The Pantiles	95 A2	Hamilton Rd	96 A1	Red Hill	60 B3	Elm Gro	98 A2
The Shaw	95 B2	Hanover Clo	96 B2	Redhouse Gdns	60 A3	Essex Rd	98 A2
Thomas St	95 A1	Havelock Rd	96 A2	The Brucks	60 B4	Ethelbert Sq	98 A2
Trinity Clo	95 B2	Hawksdown	96 A3	The Orpines	60 B3	Ethelbert Ter	98 A2
Tudor Ct	95 A3	Hawksdown Rd	96 A3	Tonbridge Rd	60 A3	Ethelred Rd	98 A2
Tunnel Rd	95 B1	Hawkshill Rd	96 B3	Upper Mill	60 A3	Fitzmary Av	98 B2
Upper Cumberland Wk	95 B3	Haywards Clo	96 A1	Warden Mill Clo	60 B4	Fulham Av	98 B3
Upper Grosvenor Rd	95 B1	Herschell Rd East	96 B1			Garrard Av	98 B2
Vale Av	95 A2	Herschell Rd West	96 B2			Glebe Gdns	98 B3
Vale Rd	95 A2	Herschell Sq	96 B2	**WESTERHAM**		Glebe Rd	98 B3
Varney St	95 B1	Hillcrest Gdns	96 A2	Ash Rd	97 B1	Gordon Dri	98 A2
Vernon Rd	95 B1	James Hall Gdns	96 A1	Bartlett Rd	97 A2	Gresham Av	98 B2
Victoria Rd	95 B1	John Tapping Clo	96 A3	Beggars La	97 B1	Guildford Av	98 A2
Wallace Clo	95 A3	Jubilee Dri	96 B1	Bloomfield Ter	97 B1	Harold Av	98 A2
Warwick Pk	95 A2	Kelvedon Rd	96 B2	Buckham Thorns Rd	97 B2	High St	98 B3
Warwick Rd	95 A2	Kennedy Dri	96 A2	Costells Meadow	97 B2	Hockeredge Gdns	98 A2
Western Rd	95 B1	King St	96 B2	Croft Rd	97 A1	Ivanhoe Rd	98 A2
William St	95 A1	Kingsdown Rd	96 B2	Croydon Rd	97 A1	Kingfisher Clo	98 B3
Willow Tree Rd	95 A3	Kingsland Gdns	96 A3	Darenth Gdns	97 B2	Kingston Av	98 B3
Windmill St	95 B2	Knoll Pl	96 B2	Delagarde Rd	97 A2	Langham Clo	98 B2
Wood St	95 B1	La Tene	96 A2	Elm Rd	97 B1	Lenham Gdns	98 B3
Woodbury Park Gdns	95 B1	Lambert Ho	96 A2	Farley Croft	97 A2	Leslie Av	98 B3
Woodbury Park Rd	95 A1	Lawn Rd	96 B2	Farley La	97 A2	Leybourne Dri	98 B2
Wybourne Rise	95 B3	Leas Rd	96 A1	Fullers Hill	97 B2	Linden Clo	98 A3
York Rd	95 A2	Liverpool Rd	96 B3	Goodley Stock	97 A3	Linden Rd	98 A3
		Lord Wardens Av	96 B2	Grange Clo	97 A2	Linksfield Rd	98 A3
UPCHURCH		Lydia Rd	96 A2	Granville Rd	97 A2	Lymington Rd	98 A3
Bishop La	93 B4	Manor Av	96 A1	Hartley Rd	97 B1	Maynard Av	98 B2
Bradshaw Clo	93 B3	Manor Rd	96 A1	High St	97 A3	Meadow Rd	98 B2
Canterbury La	93 A4	Marine Rd	96 B1	Hollingworth Way	97 B2	Michelle Gdns	98 B2
Chaffes La	93 B4	Mayers Rd	96 A3	Hortons Way	97 B2	Minster Rd	98 A2
Church Farm Rd	93 B3	Menzies Av	96 A3	Hosey Common Rd	97 B2	Mordaunt Av	98 A2
Crosier Ct	93 B3	Meryl Gdns	96 A2	Lodge La	97 A2	Mutrix Gdns	98 B2
Drakes Clo	93 B4	Mill Hill	96 A2	London Rd	97 B2	Mutrix Rd	98 B2
Forges La	93 B3	Mill Rd	96 A1	Madam Rd	97 B1	Noble Clo	98 B2
Horsham Hill	93 B3	Milldale Clo	96 A1	Market Sq	97 A2	Noble Gdns	98 B2
Horsham La	93 A3	Nevill Gdns	96 A3	Market Way	97 B2	Norman Rd	98 A2
Lower Rainham Rd	93 A4	Newlands	96 A3	Marwell	97 A2	Oaktree Gro	98 B2
Marstan Clo	93 B4	North Barrack Rd	96 B1	Mill La	97 A3	Old Boundary Rd	98 A2
Oak La	93 B4	Owen Sq	96 A2	Mill St	97 B2	Old Cressing Rd	98 B2
Otterham Quay La	93 A4	Palmerston Av	96 B2	New St	97 A2	Orchard Gdns	98 B2
The Poles	93 B3	Palmerston Ct	96 B2	Nursery Side	97 A2	Orchard Rd	98 B2
The Street	93 B3	Pilots Av	96 A1	Oak Rd	97 B1	Osbourn Av	98 A2
Wallbridge La	93 A3	Pittock Ho	96 A1	Quebec Av	97 B2	Pembroke Av	98 A3
		Poets Walk	96 A3	Railway Ter	97 B1	Prospect Clo	98 A3
WALMER		Quern Rd	96 A2	Rysted La	97 A2	Queens Rd	98 A2
Addelam Rd	96 A1	Reading Clo	96 A3	Sandy La	97 B1	Quex Rd	98 A2
Alexandra Rd	96 B2	Redsull Av	96 A1	South Bank	97 B2	Reculvers Rd	98 A3
Archery Sq Nth	96 B2	Roselands	96 A3	Squerryes Mede	97 A2	Redhill Rd	98 A3
Archery Sq Sth	96 B2	St Clare Rd	96 A3	Stratton Ter	97 A2	Richborough Rd	98 A3
Arthur Rd	96 A1	St Leonards Clo	96 A1	The Flyers Way	97 B1	Roselawn Gdns	98 B2
Balfour Rd	96 B2	St Leonards Rd	96 A1	The Green	97 A2	Rowena Rd	98 A2
Beauchamp Av	96 A1	St Marys Rd	96 B3	The Paddock	97 A2	Roxburgh Rd	98 B2
Belmont	96 A3	St Mildreds Ct	96 B3	Trotts La	97 A2	Royal Esplanade	98 B2
Blenheim Av	96 B1	St Richards Rd	96 A2	Vicarage Hill	97 B2	St Benets Rd	98 A3
Bruce Clo	96 A1	Salisbury Rd	96 A2	Wellers Clo	97 A2	St Clements Rd	98 A3
Cambridge Rd	96 B1	Shaftesbury Ct	96 B2	Westbury Ter	97 A2	St Crispins Rd	98 A3
Campbell Rd	96 B1	Somerset Rd	96 A2	Westways	97 A2	St James Park Rd	98 B2
Canada Rd	96 B1	Station App	96 A3			St Jeans Rd	98 A3
Castalia Cotts	96 B2	Station Rd	96 A3			St Lukes Rd	98 A3
Celtic Rd	96 A2	Station Rd Clo	96 A3	**WESTGATE-ON-SEA**		St Margarets Rd	98 A3
Channel Lea	96 B3	Stockdale Gdns	96 A1	Adrian Mews	98 A2	St Mildreds Gdns	98 A2
Chapman Ho	96 A1	Stoney Path	96 B1	Adrian Sq	98 A2	St Mildreds Rd	98 A2
Charles Ho	96 A1	Sydney Rd	96 A3	Alicia Av	98 B2	Saxon Rd	98 A2
Charles Rd	96 A1	Telegraph Rd	96 A2	Ascot Gdns	98 A3	Sea Rd	98 A2
Cheriton Pl	96 B1	The Beach	96 B1	Audley Av	98 B2	Seymour Av	98 B2
Cheriton Rd	96 B1	The Shrubbery	96 A3	Balmoral Rd	98 B2	Southwood Pl	98 A2
Church Path, Upper Deal	96 A1	The Strand	96 B1	Barn Cres	98 B2	Station Rd	98 A2
Church Path, Upper Walmer	96 A3	Thompson Clo	96 A3	Barnes Av	98 A2	Stephens Clo	98 B3
Church St	96 A2	Thornbridge Rd	96 A2	Beach Av	98 A2	Streete Ct	98 A2
Churchill Av	96 B2	Trinity Pl	96 A1	Beach Rise	98 A2	Streete Court Rd	98 A3
Clarence Rd	96 B2	Walmer Castle Rd	96 B2	Belmont Rd	98 A3	Sudbury Pl	98 A3
Clifford Gdns	96 A2	Walmer Way	96 A2	Birds Av	98 B3	Suffolk Pl	98 A3
Cornwall Rd	96 B1	Warwick Rd	96 B2	Boleyn Av	98 B2	Sussex Gdns	98 A2
Court Rd	96 A3	Wellesley Av	96 B2	Bowes Av	98 B2	Thanet Rd	98 A2
Cowdray Sq	96 A1	Wellington Par	96 B2	Briary Clo	98 B2	The Courts	98 B2
Curzon Clo	96 B2	Whiteacre Dri	96 A3	Bridge Rd	98 B2	The Grove	98 A2
Davis Av	96 A1	Willingdon Pl	96 B2	Brooke Av	98 B3	The School Clo	98 A2
Devon Av	96 B2	Woodstock Av	96 B1	Cambourne Av	98 A3	Tyson Av	98 B2
Dorset Ct	96 B2	Woolastone Rd	96 B1	Camellia Clo	98 A3	Vestey Ct	98 A2
Dorset Gdns	96 A2	York & Albany Clo	96 B1	Canterbury Rd	98 A3	Victoria Av	98 A3
Dover Rd	96 A3	York Rd	96 B1	Cedric Rd	98 A2	Waterside Dri	98 A3
Downlands	96 A3			Chester Rd	98 A2	Wellesley Clo	98 A3
Downs Rd	96 A2	**WATERINGBURY**		Cliffe Av	98 B2	Wellesley Rd	98 A3
Edgar Ho	96 A1	Allington Gdns	60 B3	Coronation Cres	98 B2	Wellington Clo	98 A2
Forelands Sq	96 A2	Bishops Clo	60 A4	Courtlands Way	98 A2	Wellington Rd	98 A3
Frederick Rd	96 A1	Bow Rd	60 A4	Crofton Rd	98 A3	Welsdene Rd	98 B3
Freemans Way	96 A1	Bryant Clo	60 A4	Crow Hill Rd	98 B2	Wentworth Av	98 A2
Gaunts Clo	96 A1	Canon La	60 A3	Cuthbert Rd	98 A2	Westbrook Av	98 B2
Gilham Gro	96 A1	Cobbs Clo	60 B3	Dent De Lion Rd	98 A2	Westbury Rd	98 A2
		Fields La	60 B4	Dent De Lion Rd,		Westcliff Gdns	98 B2
		Glebe Meadows	60 B3			Westgate Bay Av	98 A2
						Westonville Av	98 B2

WEST KINGSDOWN

Ash Tree Clo	99 B2
Ash Tree Dri	99 B2
Astor Rd	99 A2
Bakers Av	99 A2
Birch Way	99 B2
Blackthorn Clo	99 B3
Blue Chalet Ind Park	99 A1
Botsom La	99 A2
Church Rd	99 A2
Clearways Ind Est	99 A2
Crowhurst La	99 B3
Fawkham Rd	99 B3
Forge La	99 B3
Gillies Rd	99 A1
Gorse Hill	99 A1
Hever Av	99 A2
Hever Rd	99 A2
Hever Wood Rd	99 A2
Hollywood La	99 B3
Howells Clo	99 A2
Kingsingfield Clo	99 A2
Kingsingfield Rd	99 A2
Knatts Valley Rd	99 A2
London Rd	99 A1
Lovelace Clo	99 A2
Meadow Bank Clo	99 B2
Millfield Rd	99 A2
Mitchem Clo	99 A2
Multon Rd	99 A2
Neal Rd	99 A2
Oaklands Clo	99 A2
Pells La	99 B3
Penshurst Clo	99 A2
Phelps Clo	99 A1
Pound Bank Clo	99 B3
Regency Clo	99 A2
Rogers Wood Rd	99 B1
Rushetts Rd	99 A2
School La	99 B3
Sherbourne Clo	99 A2
Southfields Rd	99 B2
Stacklands Clo	99 A1
Symonds Clo	99 A1
The Briars	99 A2
The Grange	99 B3
The Grove	99 B3
Vernon Clo	99 B3
Viking Way	99 A1
Warland Rd	99 B2
Warland Rd E	99 B2
West Kingsdown Ind Est	99 B3
Whitegate Av	99 A2
Wood View Clo	99 A2

WEST MALLING

Alma Rd	100 A2
Birling Rd	100 B1
Brickfields	100 B2
Church Fields	100 B2
Church Rd	100 A1
County Gro	100 B2
Epsom Clo	100 A2
Ewell Av	100 A2
Fartherwell Av	100 A2
Fartherwell La	100 A2
Frog La	100 B2
Grange Clo	100 B1
High St	100 B2
King Hill	100 A3
Kings St	100 B2
Lavenders Rd	100 B3
London Rd	100 A1
Mairs Nest	100 B2
Mill Yd	100 B2
Norman Rd	100 B2
Offham Rd	100 A3
Old Ryarsh La	100 A2
Park Rd	100 B1
Police Station Rd	100 B2
Roughetts La	100 A1
Ryarsh La	100 B2
St Leonards St	100 A3
Sandown Rd	100 A2
Sandy La	100 A2
Stratford Rd	100 A2
Swan St	100 B2
Teston Rd	100 A3
Town Hill	100 B2
Town Hill Clo	100 B2
Water La	100 B2
West St	100 B2
Wickens Pl	100 B2
Windmill La	100 B3
Woodland Clo	100 A2

WHITSTABLE

Acton Rd	101 A1
Albert St	101 A1
Alexandra Rd	101 A2
All Saints Clo	101 B2
Argyle Rd	101 A2
Athol Rd	101 B1
Avondale Clo	101 B2
Baddlesmere Rd	101 B1
Baliol Rd	101 A1
Bayview Rd	101 A3
Beach Walk	101 A1
Bellevue Rd	101 B2
Belmont Rd	101 A2
Belton Clo	101 A2
Benacre Rd	101 A3
Bennells Av	101 B1
Beresford Rd	101 A1
Bexley St	101 A1
Birch Rd	101 B2
Borstal Hill	101 A2
Bridewell Pk	101 B2
Canterbury Rd	101 A2
Castle Rd	101 A1
Church St	101 B2
Clapham Hill	101 A3
Clare Rd	101 A1
Cliff Rd	101 B1
Clifford Rd	101 B2
Clifton Rd	101 A2
Clovelly Rd	101 A3
Collingwood Rd	101 A2
Cornwallis Circle	101 A2
Cranleigh Gdns	101 A2
Cromwell Rd	101 A1
Doggeral Acre	101 B3
Douglas Av	101 A2
Downs Av	101 A2
Duncan Rd	101 A2
Ellis Rd	101 B1
Enticott Clo	101 B2
Essex St	101 A2
Farmhouse Clo	101 B2
Firbanks	101 B2
Fitzroy Rd	101 B1
Forge La	101 A2
Fountain St	101 A1
Foxgrove Rd	101 B1
Friars Clo	101 B1
Gann Rd	101 B1
Gladstone Rd	101 A1
Glebe Way	101 A2
Glenside	101 B2
Gloucester Rd	101 B1
Golden Hill	101 B3
Gordon Rd	101 A2
Gorrell Rd	101 A2
Gosselin St	101 A2
Grasmere Rd	101 B2
Graystone Rd	101 B1
Green St	101 A2
Grimshill Rd	101 A2
Grosvenor Rd	101 A2
Ham Shades La	101 B2
Hamilton Rd	101 A1
Harbour St	101 A1
Harwich St	101 A2
High St	101 A1
Hill View Rd	101 A2
Hillside Rd	101 B2
Horsebridge Rd	101 A1
Hunters Chase	101 A3
Invicta Rd	101 B2
Irish Village	101 A2
Island Wall	101 A1
Ivyhouse Rd	101 B2
Joy La	101 A2
Juniper Clo	101 B2
Kent St	101 A2
King Edward St	101 A1
Kings Av	101 A2
Kingsdown Pk	101 B1
Kingsley Rd	101 A2
Linden Av	101 B2
Manor Rd	101 B1
Marine Par	101 B1
Maugham Ct	101 A2
Middle Wall	101 A1
Millfield Manor	101 A2
Millstream Clo	101 A2
Millstrood Rd	101 A2
Montpelier Av	101 A3
Nacholt Clo	101 B1
Nelson Rd	101 A2
Norfolk St	101 A2
Norman Rd	101 A2
Northwood Rd	101 A1
Nursery Clo	101 B2
Oakwood Dri	101 B1
Old Bridge Rd	101 A2
Old Farm Clo	101 A3
Orchard Clo	101 B1
Oxford St	101 A2
Paddock Vw	101 A3
Park Av	101 B1
Pier Av	101 B1
Pierpoint Rd	101 A3
Pine Tree Clo	101 B1
Queens Rd	101 B1
Railway Av	101 A1
Rayham Rd	101 B2
Regency Clo	101 B2
Regent St	101 A1
Reservoir Rd	101 A1
Rosemary Gdns	101 B2
Saddleton Rd	101 A2
St Andrews Clo	101 A3
St Annes Rd	101 B1
St Davids Clo	101 A2
St James Gdns	101 A2
St Marks Clo	101 A2
St Peters St	101 A1
Salisbury Rd	101 A2
Sea St	101 A1
Sea Wall	101 A1
Seeshill Clo	101 A2
Seymour Av	101 A2
Shaftesbury Rd	101 A1
South St	101 B2
South View Rd	101 A3
Spire Av	101 B2
Spring Walk	101 A3
Stanley Rd	101 A2
Station Rd	101 A2
Strangford Rd	101 B1
Stream Walk	101 A2
Suffolk St	101 A2
Summerfield Av	101 B2
Swanfield Rd	101 A2
Sydenham St	101 A1
Sydney Rd	101 A2
Tankerton Circus	101 B1
Tankerton Mews	101 A1
Tankerton Rd	101 A1
Terrys La	101 A1
Teynham Dri	101 B1
Teynham Rd	101 A1
Thanet Way	101 A3
The Bridge App	101 B1
The Halt	101 B2
The Saltings	101 A1
The Warren	101 A3
Thurston Pk	101 A2
Toll Gate Clo	101 A2
Tower Hill	101 A1
Tower Par	101 A1
Tower Rd	101 A1
Vale Rd	101 A2
Victoria St	101 A1
Virginia Rd	101 B2
Vulcan Clo	101 A2
Walmer Rd	101 A2
Warwick Rd	101 A1
Waterloo Rd	101 A1
West Cliff	101 A2
Westgate Ter	101 A2
Westmeads Rd	101 A1
Wheatley Rd	101 A1
Windmill Rd	101 A3
Woodlawn St	101 A1
Wynn Rd	101 B1

WOODCHURCH

Appledore Rd	102 B2
Bourney Pl	102 B1
Brattle	102 B2
Cherry Orchard	102 A1
Front Rd	102 A1
Hamstreet Rd	102 B2
Kirkwood Av	102 A1
Lower Rd	102 B2
Mill Vw	102 A1
Place La	102 B1
Plum Tree Gdns	102 B2
Rectory Clo	102 A1
Susans Hill	102 A1
Tenterden Rd	102 A2
The Green	102 A1

WYE

Abbots Walk	103 A1
Bramble Clo	103 A1
Bramble La	103 A1
Bridge St	103 A1
Chequers Pk	103 A2
Cherry Garden Cres	103 B2
Cherry Garden La	103 B2
Church St	103 B2
Churchfield Way	103 A1
Coldharbour La	103 B2
Gregory Ct	103 A1
Harville Rd	103 A1
High St	103 B1
Little Chequers	103 A2
Luckley Ho	103 A2
Olantigh Rd	103 B1
Orchard Dri	103 B2
Oxenturn Rd	103 B2
Scotton St	103 B1
Taylors Yard	103 A1
The Close	103 A2
The Forstal	103 A1
The Green	103 B1
Upper Bridge St	103 B2

YALDING

Acton Pl	103 B3
Benover Rd	103 B4
Blunden La	103 B3
Downs Rd	103 B3
Hampstead La	103 A3
High St	103 B4
Kenward Rd	103 A3
Lees Rd	103 A4
Lughorse La	103 B3
Lyngs Clo	103 B4
Medway Av	103 B3
Mount Av	103 B3
Oast Ct	103 B3
The Nook	103 B3
Vicarage Rd	103 B3
Walnut Tree Cotts	103 B3
Yalding Hill	103 B3